# Student's Solutions Manual

to accompany

# Elementary Statistics
## *A Step by Step Approach*
### Seventh Edition

## Allan G. Bluman
*Professor Emeritus*
*Community College of Allegheny County*

Prepared by
## Sally Robinson
*South Plains College*

 **Higher Education**

Boston   Burr Ridge, IL   Dubuque, IA   New York   San Francisco   St. Louis
Bangkok   Bogotá   Caracas   Kuala Lumpur   Lisbon   London   Madrid   Mexico City
Milan   Montreal   New Delhi   Santiago   Seoul   Singapore   Sydney   Taipei   Toronto

The McGraw·Hill Companies

 **Higher Education**

*Student's Solutions Manual* to accompany
ELEMENTARY STATISTICS: A STEP BY STEP APPROACH, SEVENTH EDITION
BLUMAN, ALLAN

Published by McGraw-Hill Higher Education, an imprint of The McGraw-Hill Companies, Inc., 1221 Avenue of the Americas, New York, NY 10020. Copyright © 2009, 2007, 2004 by The McGraw-Hill Companies, Inc. All rights reserved.

1 2 3 4 5 6 7 8 9 0 QPD/QPD 0 9 8

ISBN: 978-0-07-333128-7
MHID: 0-07-333128-7

www.mhhe.com

# Contents

# Preface

This manual includes solutions to odd and selected even exercises in *Elementary Statistics: A Step-by-Step Approach 7e* by Allan G. Bluman. Solutions are worked out step-by-step where appropriate and generally follow the same procedures used in the examples in the textbook. Answers may be carried to several decimal places to increase accuracy and to facilitate checking. See your instructor for specific rounding rules. Graphs are included with the solutions when appropriate or required. They are intended to convey a general idea and may not be to scale.

Caution: Answers generated using graphing calculators such as the TI-83 may vary from those shown in this manual.

To maximize the assistance provided in this manual, you should:

1. Read each section of the text carefully, noting examples and formulas.

2. Begin working the exercises using textbook examples and class notes as your guide, then refer to the answers in this manual.

3. Many instructors require students to interpret their answers within the context of the problem. These interpretations are powerful tools to understanding the meaning and purpose of each calculation. You should attempt to interpret each calculation even if you are not required to do so.

4. Be sure to show your work. When checking your work for errors you will need to review each step. When preparing for exams, reviewing each step helps you to recall the process involved in producing each calculation.

5. As you gain confidence and understanding, you should attempt to work exercises without referring to examples or notes. Check each answer in the solutions manual before beginning the next exercise.

6. Slight variations between your answers and the answers in this manual are probably due to rounding differences and should not be a cause for concern. If you are concerned about these variations, check each step of your calculation again.

7. Many errors can be traced to the improper application of the rules for order of operations. You should first attempt to determine where and how your error occurred because diagnosing your error increases understanding and prevents future errors. See your instructor if you are unsure of the location or cause of your error.

Sally H. Robinson

# Chapter 1 - The Nature of Probability and Statistics

REVIEW EXERCISES - CHAPTER 1

1. Descriptive statistics describes a set of data. Inferential statistics uses a set of data to make predictions about a population.

3. Answers will vary.

5. When the population is large, the researcher saves time and money using samples. Samples are used when the units must be destroyed.

6.
   a. inferential   e. inferential
   b. descriptive   f. inferential
   c. descriptive   g. descriptive
   d. descriptive   h. inferential

7.
   a. ratio      f. ordinal
   b. ordinal    g. ratio
   c. ratio      h. ratio
   d. interval   i. nominal
   e. ratio      j. ratio

8.
   a. quantitative   e. qualitative
   b. qualitative    f. quantitative
   c. quantitative   g. qualitative
   d. quantitative

9.
   a. discrete      d. continuous
   b. continuous    e. discrete
   c. continuous    f. continuous

11. Random samples are selected by using chance methods or random numbers. Systematic samples are selected by numbering each subject and selecting every $k$th number. Stratified samples are selected by dividing the population into groups and selecting from each group. Cluster samples are selected by using intact groups called clusters.

12.
   a. cluster       d. systematic
   b. systematic    e. stratified
   c. random

13. Answers will vary.

15. Answers will vary.

17.
   a. experimental   c. observational
   b. observational  d. experimental

19. Answers will vary. Possible answers include:
(a) overall health of participants, amount of exposure to infected individuals through the workplace or home
(b) gender and/or age of driver, time of day
(c) diet, general health, heredity factors
(d) amount of exercise, heredity factors

21. Claims can be proven only if the entire population is used.

23. Since the results are not typical, the advertisers selected only a few people for whom the product worked extremely well.

25. "74% more calories" than what? No comparison group is stated.

27. What is meant by "24 hours of acid control"?

29. Possible reasons for conflicting results: The amount of caffeine in the coffee or tea or the brewing method.

31. Answers will vary.

CHAPTER QUIZ
1. True
2. False, it is a data value.
3. False, the highest level is ratio.
4. False, it is stratified sampling.
5. False, it is a quantitative variable.
6. True
7. False, it is 5.5-6.5 inches.
8. c.
9. b.
10. d.
11. a.
12. c.
13. a.
14. descriptive, inferential
15. gambling, insurance
16. population
17. sample

1

18.
a. saves time
b. saves money
c. use when population is infinite

19.
  a. random      c. cluster
  b. systematic    d. stratified

20. quasi-experimental

21. random

22.
  a. descriptive    d. inferential
  b. inferential     e. inferential
  c. descriptive

23.
  a. nominal    d. interval
  b. ratio       e. ratio
  c. ordinal

24.
  a. continuous    d. continuous
  b. discrete      e. discrete
  c. continuous

25.
  a. $47.5 - 48.5$ seconds     d. $13.65 - 13.75$ pounds
  b. $0.555 - 0.565$ centimeters   e. $6.5 - 7.5$ feet
  c. $9.05 - 9.15$ quarts

EXERCISE SET 2-1

1. Frequency distributions are used to organize data in a meaningful way, to facilitate computational procedures for statistics, to make it easier to draw charts and graphs, and to make comparisons among different sets of data.

3.
a. $11.5 - 18.5$, $\frac{12+18}{2} = \frac{30}{2} = 15$, $18.5 - 11.5 = 7$

b. $55.5 - 74.5$, $\frac{56+74}{2} = \frac{130}{2} = 65$, $74.5 - 55.5 = 19$

c. $694.5 - 705.5$, $\frac{695+705}{2} = \frac{1400}{2} = 700$, $705.5 - 694.5 = 11$

d. $13.55 - 14.75$, $\frac{13.6+14.7}{2} = \frac{28.3}{2} = 14.15$, $14.75 - 13.55 = 1.2$

e. $2.145 - 3.935$, $\frac{2.15+3.93}{2} = \frac{6.08}{2} = 3.04$, $3.935 - 2.145 = 1.79$

5.
a. Class width is not uniform.
b. Class limits overlap, and class width is not uniform.
c. A class has been omitted.
d. Class width is not uniform.

7.

| Class | Tally | f | Percent |
|---|---|---|---|
| A | IIII | 4 | 10% |
| M | HHI HHI HHI HHI HHI III | 28 | 70% |
| H | HHI I | 6 | 15% |
| S | II | 2 | 5% |
| | | 40 | 100% |

9. H = 325    L = 165
Range = 325 − 165 = 160
Width = 160 ÷ 8 = 20  round up to 21

| Limits | Boundaries | f |
|---|---|---|
| 165 - 185 | 164.5 - 185.5 | 4 |
| 186 - 206 | 185.5 - 206.5 | 6 |
| 207 - 227 | 206.5 - 227.5 | 15 |
| 228 - 248 | 227.5 - 248.5 | 13 |
| 249 - 269 | 248.5 - 269.5 | 9 |
| 270 - 290 | 269.5 - 290.5 | 1 |
| 291 - 311 | 290.5 - 311.5 | 1 |
| 312 - 332 | 311.5 - 332.5 | 1 |
| | | 50 |

9. continued
A peak occurs in class 207 - 227. There are no empty classes. Each of the three highest classes has one data value.

| | cf |
|---|---|
| Less than 164.5 | 0 |
| Less than 185.5 | 4 |
| Less than 206.5 | 10 |
| Less than 227.5 | 25 |
| Less than 248.5 | 38 |
| Less than 269.5 | 47 |
| Less than 290.5 | 48 |
| Less than 311.5 | 49 |
| Less than 332.5 | 50 |

11. H = 780    L = 746
Range = 780 − 746 = 34
Width = 34 ÷ 5 = 6.8   round up to 7

| Limits | Boundaries | f |
|---|---|---|
| 746 - 752 | 745.5 - 752.5 | 4 |
| 753 - 759 | 752.5 - 759.5 | 6 |
| 760 - 766 | 759.5 - 766.5 | 8 |
| 767 - 773 | 766.5 - 773.5 | 9 |
| 774 - 780 | 773.5 - 780.5 | 3 |
| | | 30 |

| | cf |
|---|---|
| Less than 745.5 | 0 |
| Less than 752.5 | 4 |
| Less than 759.5 | 10 |
| Less than 766.5 | 18 |
| Less than 773.5 | 27 |
| Less than 780.5 | 30 |

13. H = 70    L = 27
Range = 70 − 27 = 43
Width = 43 ÷ 7 = 6.1 or 7

| Limits | Boundaries | f |
|---|---|---|
| 27 - 33 | 26.5 - 33.5 | 7 |
| 34 - 40 | 33.5 - 40.5 | 14 |
| 41 - 47 | 40.5 - 47.5 | 15 |
| 48 - 54 | 47.5 - 54.5 | 11 |
| 55 - 61 | 54.5 - 61.5 | 3 |
| 62 - 68 | 61.5 - 68.5 | 3 |
| 69 - 75 | 68.5 - 75.5 | 2 |
| | | 55 |

13. continued

| | cf |
|---|---|
| Less than 26.5 | 0 |
| Less than 33.5 | 7 |
| Less than 40.5 | 21 |
| Less than 47.5 | 36 |
| Less than 54.5 | 47 |
| Less than 61.5 | 50 |
| Less than 68.5 | 53 |
| Less than 75.5 | 55 |

15. H = 632   L = 6

Width = 632 − 6 = 626

Range = 626 ÷ 5 = 125.2  round up to 127

| Limits | Boundaries | f |
|---|---|---|
| 6 - 132 | 5.5 - 132.5 | 16 |
| 133 - 259 | 132.5 - 259.5 | 3 |
| 260 - 386 | 259.5 - 386.5 | 0 |
| 387 - 513 | 386.5 - 513.5 | 0 |
| 514 - 640 | 513.5 - 640.5 | 1 |
| | | 20 |

The greatest concentration of data values is in the lowest class.  All but one of the data values are in the lowest two classes.  There is one extremely large data value occurring in the highest class.

| | cf |
|---|---|
| Less than 5.5 | 0 |
| Less than 132.5 | 16 |
| Less than 259.5 | 19 |
| Less than 386.5 | 19 |
| Less than 513.5 | 19 |
| Less than 640.5 | 20 |

17. H = 11,413   L = 150

Range = 11,413 − 150 = 11,263

Width = 11,263 ÷ 10 = 1126.3 or 1127

17. continued

| Limits | Boundaries | f |
|---|---|---|
| 150 - 1276 | 149.5 - 1276.5 | 2 |
| 1277 - 2403 | 1276.5 - 2403.5 | 2 |
| 2404 - 3530 | 2403.5 - 3530.5 | 5 |
| 3531 - 4657 | 3530.5 - 4657.5 | 8 |
| 4658 - 5784 | 4657.5 - 5784.5 | 7 |
| 5785 - 6911 | 5784.5 - 6911.5 | 3 |
| 6912 - 8038 | 6911.5 - 8038.5 | 7 |
| 8039 - 9165 | 8038.5 - 9165.5 | 3 |
| 9166 - 10,292 | 9165.5 - 10,292.5 | 3 |
| 10,293 - 11,419 | 10,292.5 - 11,419.5 | 2 |
| | | 42 |

| | cf |
|---|---|
| Less than 149.5 | 0 |
| Less than 1276.5 | 2 |
| Less than 2403.5 | 4 |
| Less than 3530.5 | 9 |
| Less than 4657.5 | 17 |
| Less than 5784.5 | 24 |
| Less than 6911.5 | 27 |
| Less than 8038.5 | 34 |
| Less than 9165.5 | 37 |
| Less than 10,292.5 | 40 |
| Less than 11,419.5 | 42 |

19. The percents add up to 101%.  They should total 100% unless rounding was used.

EXERCISE SET 2-2

1.

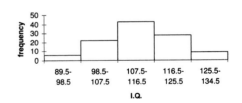

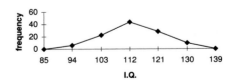

## 1. continued

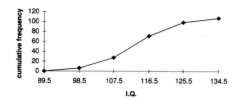

Eighty applicants do not need to enroll in the summer programs.

## 3.

| Limits | Boundaries | f |
|--------|-----------|---|
| 3 - 45 | 2.5 - 45.5 | 19 |
| 46 - 88 | 45.5 - 88.5 | 19 |
| 89 - 131 | 88.5 - 131.5 | 10 |
| 132 - 174 | 131.5 - 174.5 | 1 |
| 175 - 217 | 174.5 - 217.5 | 0 |
| 218 - 260 | 217.5 - 260.5 | 1 |
| | | 50 |

| | cf |
|--|----|
| Less than 2.5 | 0 |
| Less than 45.5 | 19 |
| Less than 88.5 | 38 |
| Less than 131.5 | 48 |
| Less than 174.5 | 49 |
| Less than 217.5 | 49 |
| Less than 260.5 | 50 |

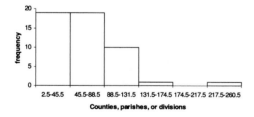

The distribution is positively skewed.

## 3. continued

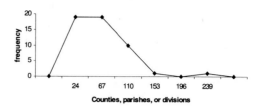

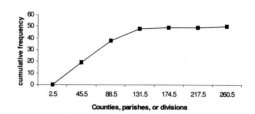

## 5.

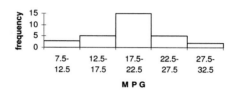

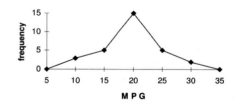

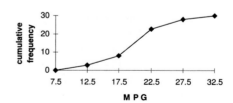

7.

| Limits | Boundaries | $f$ (1998) | $f$ (2003) |
|--------|-----------|-----------|-----------|
| 0 - 22 | -0.5 - 22.5 | 18 | 26 |
| 23 - 45 | 22.5 - 45.5 | 7 | 1 |
| 46 - 68 | 45.5 - 68.5 | 3 | 0 |
| 69 - 91 | 68.5 - 91.5 | 1 | 1 |
| 92 - 114 | 91.5 - 114.5 | 1 | 0 |
| 115 - 137 | 114.5 - 137.5 | 0 | 1 |
| 138 - 160 | 137.5 - 160.5 | 0 | 1 |
| | | 30 | 30 |

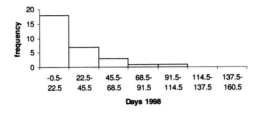

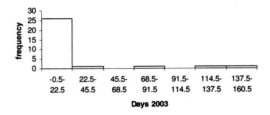

Both distributions are positively skewed, but the data are somewhat more spread out in the first three classes in 1998 than in 2003. There are two large data values in the 2003 data.

9.

| Limits | Boundaries | $f$ |
|--------|-----------|-----|
| 83.1 - 90.0 | 83.05 - 90.05 | 3 |
| 90.1 - 97.0 | 90.05 - 97.05 | 5 |
| 97.1 - 104.0 | 97.05 - 104.05 | 6 |
| 104.1 - 111.0 | 104.05 - 111.05 | 7 |
| 111.1 - 118.0 | 111.05 - 118.05 | 3 |
| 118.1 - 125.0 | 118.05 - 125.05 | 1 |
| | | 25 |

| | $cf$ |
|--------|-----|
| Less than 83.05 | 0 |
| Less than 90.05 | 3 |
| Less than 97.05 | 8 |
| Less than 104.05 | 14 |
| Less than 111.05 | 21 |
| Less than 118.05 | 24 |
| Less than 125.05 | 25 |

9. continued

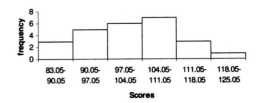

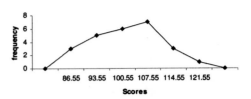

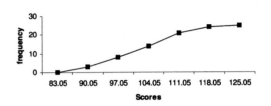

11.

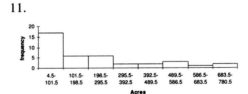

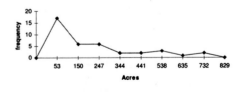

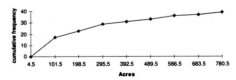

The peak is in the first class, and then the histogram is rather uniform after the first class. Most of the parks have less than 101.5 thousand acres as compared with any other class of values.

**13.**

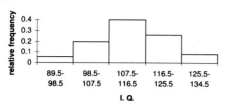

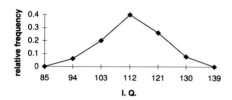

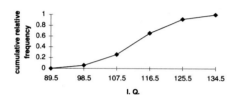

The proportion of applicants who need to enroll in a summer program is 0.26 or 26%.

15. H = 270    L = 80
Range = 270 − 80 = 190
Width = 190 ÷ 7 = 27.1 or 28
Use width = 29 (rule 2)

| Limits | Boundaries | f | rf |
|--------|-----------|---|-----|
| 80 - 108 | 79.5 - 108.5 | 8 | 0.17 |
| 109 - 137 | 108.5 - 137.5 | 13 | 0.28 |
| 138 - 166 | 137.5 - 166.5 | 2 | 0.04 |
| 167 - 195 | 166.5 - 195.5 | 9 | 0.20 |
| 196 - 224 | 195.5 - 224.5 | 10 | 0.22 |
| 225 - 253 | 224.5 - 253.5 | 2 | 0.04 |
| 254 - 282 | 253.5 - 282.5 | 2 | 0.04 |
|  |  |  | 0.99* |

*due to rounding

|  | crf |
|--|-----|
| Less than 79.5 | 0.00 |
| Less than 108.5 | 0.17 |
| Less than 137.5 | 0.45 |
| Less than 166.5 | 0.49 |
| Less than 195.5 | 0.69 |
| Less than 224.5 | 0.91 |
| Less than 253.5 | 0.95 |
| Less than 282.5 | 0.99 |

**15. continued**

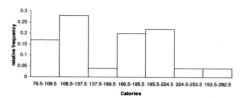

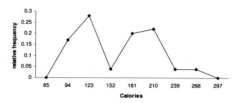

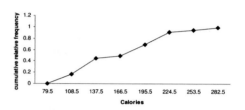

The histogram has two peaks.

17.

| Boundaries | rf | crf |
|-----------|-----|-----|
| -0.5 - 27.5 | 0.63 | 0.63 |
| 27.5 - 55.5 | 0.20 | 0.83 |
| 55.5 - 83.5 | 0.07 | 0.90 |
| 83.5 - 111.5 | 0.00 | 0.90 |
| 111.5 - 139.5 | 0.00 | 0.90 |
| 139.5 - 167.5 | 0.10 | 1.00 |
| 167.5 - 195.5 | 0.00 | 1.00 |
|  | 100.0 |  |

|  | crf |
|--|-----|
| Less than -0.5 | 0.00 |
| Less than 27.5 | 0.63 |
| Less than 55.5 | 0.83 |
| Less than 83.5 | 0.90 |
| Less than 111.5 | 0.90 |
| Less than 139.5 | 0.90 |
| Less than 167.5 | 1.00 |

17. continued

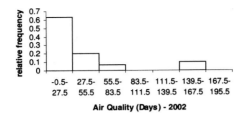

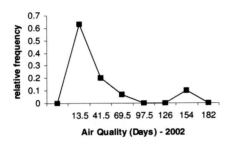

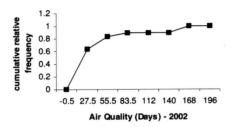

19.

| Limits | Boundaries | $X_m$ | f |
|---|---|---|---|
| 22 - 24 | 21.5 - 24.5 | 23 | 1 |
| 25 - 27 | 24.5 - 27.5 | 26 | 3 |
| 28 - 30 | 27.5 - 30.5 | 29 | 0 |
| 31 - 33 | 30.5 - 33.5 | 32 | 6 |
| 34 - 36 | 33.5 - 36.5 | 35 | 5 |
| 37 - 39 | 36.5 - 39.5 | 38 | 3 |
| 40 - 42 | 39.5 - 42.5 | 41 | 2 |
| | | | 20 |

| | cf |
|---|---|
| Less than 21.5 | 0 |
| Less than 24.5 | 1 |
| Less than 27.5 | 4 |
| Less than 30.5 | 4 |
| Less than 33.5 | 10 |
| Less than 36.5 | 15 |
| Less than 39.5 | 18 |
| Less than 42.5 | 20 |

19. continued

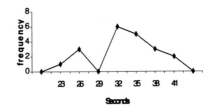

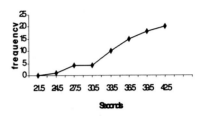

**EXERCISE SET 2-3**

1.

| | f |
|---|---|
| UCLA | 11 |
| Texas A & M | 2 |
| Calif. State | 1 |
| Arizona | 6 |
| Fresno State | 1 |
| Oklahoma | 1 |
| California | 1 |
| Michigan | 1 |
| | 24 |

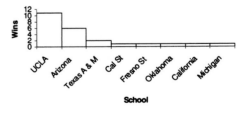

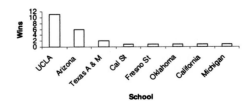

**3.**
The best place to market products would be to residential (home) users.

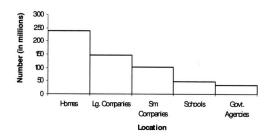

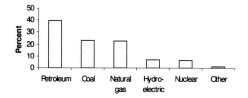

**5.**

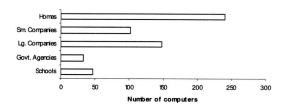

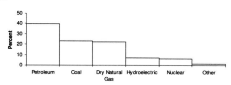

**7.**

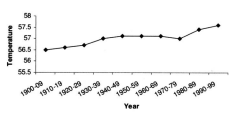

**9.**

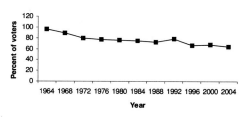

The sample used was not representative of the general population.

**11.**

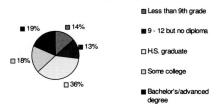

**13.**

| Career change | 34% | 122.4° |
|---|---|---|
| New job | 29% | 104.4° |
| Start business | 21% | 75.6° |
| Retire | 16% | 57.6° |
| | 100% | 360.0° |

Pie chart:

**13. continued**
Pareto chart:

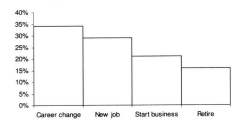

The pie graph better represents the data since we are looking at parts of a whole.

**15.**

| 4 | 2 | 3 | | | | | | | | |
|---|---|---|---|---|---|---|---|---|---|---|
| 4 | 6 | 6 | 7 | 8 | 9 | 9 | | | | |
| 5 | 0 | 1 | 1 | 1 | 1 | 2 | 2 | 4 | 4 | 4 4 |
| 5 | 5 | 5 | 5 | 5 | 6 | 6 | 6 | 7 | 7 | 7 7 8 |
| 6 | 0 | 1 | 1 | 1 | 2 | 4 | 4 | | | |
| 6 | 5 | 8 | 9 | | | | | | | |

The distribution is somewhat symmetric and unimodal. The majority of the Presidents were in their 50's when inaugurated.

**17.**

```
     Variety 1                      Variety 2
                      2 | 1 | 3  8
                  3   0 | 2 | 5
              9 8 8 5  2 | 3 | 6  8
                3 3   1 | 4 | 1  2  5  5
  9 9 8 5 3 3 2 1  0 | 5 | 0  3  5  5  6  7  9
                      6 | 2  2
```

The distributions are similar but variety 2 seems to be more variable than variety 1.

**19.**
Answers will vary.

**21.**

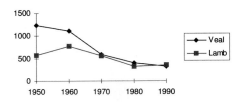

In 1950, veal production was considerably higher than lamb. By 1970, production was approximately the same for both.

**23.**

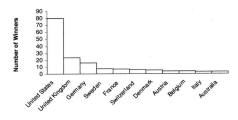

**25.** The values on the $y$ axis start at 3.5. Also there are no data values shown for the years 2004 through 2011.

**REVIEW EXERCISES - CHAPTER 2**

**1.**

| Class | f |
|---|---|
| Newspaper | 10 |
| Television | 16 |
| Radio | 12 |
| Internet | 12 |
| | 50 |

**3.**

| Class | f |
|---|---|
| baseball | 4 |
| golf ball | 5 |
| tennis ball | 6 |
| soccer ball | 5 |
| football | 5 |
| | 25 |

**5.**

| Class | f | | cf |
|---|---|---|---|
| 11 | 1 | less than 10.5 | 0 |
| 12 | 2 | less than 11.5 | 1 |
| 13 | 2 | less than 12.5 | 3 |
| 14 | 2 | less than 13.5 | 5 |
| 15 | 1 | less than 14.5 | 7 |
| 16 | 2 | less than 15.5 | 8 |
| 17 | 4 | less than 16.5 | 10 |
| 18 | 2 | less than 17.5 | 14 |
| 19 | 2 | less than 18.5 | 16 |
| 20 | 1 | less than 19.5 | 18 |
| 21 | 0 | less than 20.5 | 19 |
| 22 | 1 | less than 21.5 | 19 |
| | 20 | less than 22.5 | 20 |

**7.**

| Limits | Boundaries | f |
|--------|-----------|---|
| 15 - 19 | 14.5 - 19.5 | 3 |
| 20 - 24 | 19.5 - 24.5 | 18 |
| 25 - 29 | 24.5 - 29.5 | 18 |
| 30 - 34 | 29.5 - 34.5 | 8 |
| 35 - 39 | 34.5 - 39.5 | 3 |
| | | 50 |

| | cf |
|--|--|
| Less than 14.5 | 0 |
| Less than 19.5 | 3 |
| Less than 24.5 | 21 |
| Less than 29.5 | 39 |
| Less than 34.5 | 47 |
| Less than 39.5 | 50 |

**9.**

| Limits | Boundaries | f | | cf |
|--------|-----------|---|--|----|
| 170 - 188 | 169.5 - 188.5 | 11 | less than 169.5 | 0 |
| 189 - 207 | 188.5 - 207.5 | 9 | less than 188.5 | 11 |
| 208 - 226 | 207.5 - 226.5 | 4 | less than 207.5 | 20 |
| 227 - 245 | 226.5 - 245.5 | 5 | less than 226.5 | 24 |
| 246 - 264 | 245.5 - 264.5 | 0 | less than 245.5 | 29 |
| 265 - 283 | 264.5 - 283.5 | 0 | less than 264.5 | 29 |
| 284 - 302 | 283.5 - 302.5 | 0 | less than 283.5 | 29 |
| 303 - 321 | 302.5 - 321.5 | 1 | less than 302.5 | 29 |
| | | 30 | less than 321.5 | 30 |

**11.**

| Limits | Boundaries | rf |
|--------|-----------|----|
| 51 - 59 | 50.5 - 59.5 | 0.125 |
| 60 - 68 | 59.5 - 68.5 | 0.300 |
| 69 - 77 | 68.5 - 77.5 | 0.275 |
| 78 - 86 | 77.5 - 86.5 | 0.200 |
| 87 - 95 | 86.5 - 95.5 | 0.050 |
| 96 - 104 | 95.5 - 104.5 | 0.050 |
| | | 1.000 |

| | crf |
|--|-----|
| Less than 50.5 | 0.000 |
| Less than 59.5 | 0.125 |
| Less than 68.5 | 0.425 |
| Less than 77.5 | 0.700 |
| Less than 86.5 | 0.900 |
| Less than 95.5 | 0.950 |
| Less than 104.5 | 1.000 |

**11. continued**

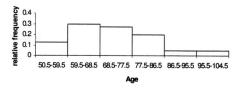

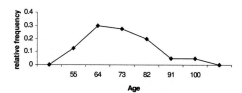

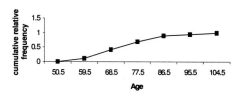

**13.**

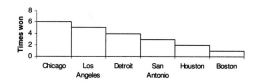

**15.**

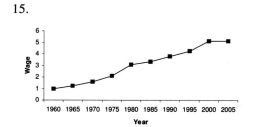

The minimum wage has increased over the years with the largest increase occurring between 1975 and 1980.

17.

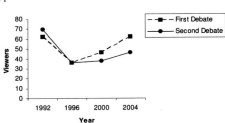

About the same number of people watched the first and second debates in 1992 and 1996. After that more people watched the first debate than watched the second debate.

19.

The majority of people surveyed would like to spend the rest of their careers with their present employer.

21.

| 10 | 2 8 8 |
|----|-------|
| 11 | 3 |
| 12 | |
| 13 | |
| 14 | 2 4 |
| 15 | |
| 16 | |
| 17 | 6 6 6 |
| 18 | 4 9 |
| 19 | 2 |
| 20 | 5 9 |
| 21 | 0 |

## CHAPTER 2 QUIZ

1. False
2. False
3. False
4. True
5. True
6. False
7. False
8. c.
9. c.
10. b.

11. b.
12. Categorical, ungrouped, grouped
13. 5, 20
14. categorical
15. time series
16. stem and leaf plot
17. vertical or y
18.

| Class | f | cf |
|-------|---|----|
| H | 6 | 6 |
| A | 5 | 11 |
| M | 6 | 17 |
| C | 8 | 25 |
| | 25 | |

19.

20.

| Class | f | | cf |
|-------|---|---|----|
| 0.5 − 1.5 | 1 | less than 0.5 | 0 |
| 1.5 − 2.5 | 5 | less than 1.5 | 1 |
| 2.5 − 3.5 | 3 | less than 2.5 | 6 |
| 3.5 − 4.5 | 4 | less than 3.5 | 9 |
| 4.5 − 5.5 | 2 | less than 4.5 | 13 |
| 5.5 − 6.5 | 6 | less than 5.5 | 15 |
| 6.5 − 7.5 | 2 | less than 6.5 | 21 |
| 7.5 − 8.5 | 3 | less than 7.5 | 23 |
| 8.5 − 9.5 | 4 | less than 8.5 | 26 |
| | 30 | less than 9.5 | 30 |

21.

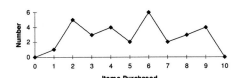

21. continued

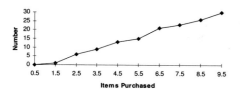

22.

| Limits | Boundaries | f |
|--------|-----------|---|
| 27 - 90 | 26.5 - 90.5 | 13 |
| 91 - 154 | 90.5 - 154.5 | 2 |
| 155 - 218 | 154.5 - 218.5 | 0 |
| 219 - 282 | 218.5 - 282.5 | 5 |
| 283 - 346 | 282.5 - 346.5 | 0 |
| 347 - 410 | 346.5 - 410.5 | 2 |
| 411 - 474 | 410.5 - 474.5 | 0 |
| 475 - 538 | 474.5 - 538.5 | 1 |
| 539 - 602 | 538.5 - 602.5 | 2 |
| | | 25 |

23.

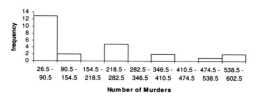

The distribution is positively skewed with one more than half of the data values in the lowest class.

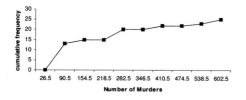

24.

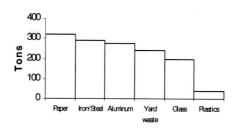

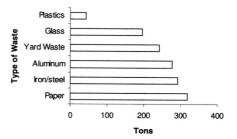

25.

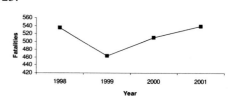

Fatalities decreased in 1999 and then increased the next two years.

26.

| 1 | 5 | 9 | | |
|---|---|---|---|---|
| 2 | 6 | 8 | | |
| 3 | 1 | 5 | 8 | 8 | 9 |
| 4 | 1 | 7 | 8 | |
| 5 | 3 | 3 | 4 | |
| 6 | 2 | 3 | 7 | 8 |
| 7 | 6 | 9 | | |
| 8 | 6 | 8 | 9 | |
| 9 | 8 | | | |

Note: Answers may vary due to rounding, TI 83's, or computer programs.

EXERCISE SET 3-1

1.
a. $\overline{X} = \frac{\Sigma X}{n} = \frac{93.09}{25} = 3.7236 \approx 3.724$

b. MD: 3.57, 3.64, 3.64, 3.65, 3.66, 3.67, 3.67, 3.68, 3.7, 3.7, 3.7, 3.73, **3.73**, 3.74, 3.74, 3.74, 3.75, 3.76, 3.77, 3.78, 3.78, 3.8, 3.8, 3.83, 3.86

c. Mode: 3.7 and 3.74    d. MR: $\frac{3.57 + 3.86}{2} = 3.715$

3.
a. $\overline{X} = \frac{\Sigma X}{n} = \frac{1566}{23} = 68.1$

b. MD: 38, 42, 42, ..., **68**, ..., 85, 90, 91

c. Mode = 42, 62, 64, 66, 72, 74    d. MR $= \frac{38 + 91}{2} = 64.5$

For the best measure of average, answers will vary.

5.
a. $\overline{X} = \frac{\Sigma X}{n} = \frac{150,755}{16} = 9422.2$

b. MD $= \frac{8632 + 9344}{2} = 8988$

c. Mode: 7552, 8632, 12568    d. MR $= \frac{6300 + 12568}{2} = 9434$

The claim seems a litle high.

7.
a. $\overline{X} = \frac{\Sigma X}{n} = \frac{79.6}{12} = 6.63$

b. MD: 5.4, 5.4, 6.2, 6.2, 6.4, **6.4, 6.5**, 7.0, 7.2, 7.2, 7.7, 8.0        MD $= \frac{6.4 + 6.5}{2} = 6.45$

c. Mode: 5.4, 6.2, 6.4, 7.2    d. MR $= \frac{5.4 + 8.0}{2} = 6.7$

For the best measure of average, answers will vary.

9.
a. $\overline{X} = \frac{\Sigma X}{n} = \frac{238,512}{42} = 5678.9$

b. MD: 150, 885, ..., **5315, 5370**, ..., 11070, 11413        MD $= \frac{5315 + 5370}{2} = 5342.5$

c. Mode: 4450        d. MR $= \frac{150 + 11,413}{2} = 5781.5$

The distribution is skewed to the right.

11.
For 2004:

a. $\overline{X} = \frac{\Sigma X}{n} = \frac{202,108}{24} = 8421.2$        b. MD $= \frac{8174 + 8220}{2} = 8197$

11. continued

c. Mode: no mode

d. $MR = \frac{5044 + 14925}{2} = 9984.5$

For 1990:

a. $\overline{X} = \frac{\Sigma X}{n} = \frac{235,440}{24} = 9810$

b. $MD = \frac{9200 + 9229}{2} = 9214.5$

Mode: no mode

d. $MR = \frac{4768 + 21923}{2} = 13,345.5$

Based on this data, it appears that the population is declining.

13.

| Class Limits | Boundaries | $X_m$ | $f$ | $f \cdot X_m$ |
|---|---|---|---|---|
| 2.48 - 7.48 | 2.475 - 7.485 | 4.98 | 7 | 34.86 |
| 7.49 - 12.49 | 7.485 - 12.495 | 9.99 | 3 | 29.97 |
| 12.50 - 17.50 | 12.495 - 17.505 | 15.00 | 1 | 15.00 |
| 17.51 - 22.51 | 17.505 - 22.515 | 20.01 | 7 | 140.07 |
| 22.52 - 27.52 | 2.515 - 27.525 | 25.02 | 5 | 125.10 |
| 27.53 - 32.53 | 27.525 - 32.535 | 30.03 | 5 | 150.15 |
| | | | 28 | 495.15 |

a. $\overline{X} = \frac{\Sigma f \cdot X_m}{n} = \frac{495.15}{28} = 17.684$

b. modal class: $2.48 - 7.48$ and $17.51 - 22.51$     The U. S. mean is less.

15.

| Percentage | Boundaries | $X_m$ | $f$ | $f \cdot X_m$ |
|---|---|---|---|---|
| 0.8 - 4.4 | 0.75 - 4.45 | 2.6 | 26 | 67.6 |
| 4.5 - 8.1 | 4.45 - 8.15 | 6.3 | 11 | 69.3 |
| 8.2 - 11.8 | 8.15 - 11.85 | 10.0 | 4 | 40.0 |
| 11.9 - 15.5 | 11.85 - 15.55 | 13.7 | 5 | 68.5 |
| 15.6 - 19.2 | 15.55 - 19.25 | 17.4 | 2 | 34.8 |
| 19.3 - 22.9 | 19.25 - 22.95 | 21.1 | 1 | 21.1 |
| 23.0 - 26.6 | 22.95 - 26.65 | 24.8 | 1 | 24.8 |
| | | | 50 | 326.1 |

a. $\overline{X} = \frac{\Sigma f \cdot X_m}{n} = \frac{326.1}{50} = 6.52$

b. modal class: $0.8 - 4.4$

The mean is probably not the best measure of central tendency for this data because the data is "top" heavy.

17.

| Percentage | Boundaries | $X_m$ | $f$ | $f \cdot X_m$ |
|---|---|---|---|---|
| 15.2 - 19.6 | 15.15 - 19.65 | 17.4 | 3 | 52.2 |
| 19.7 - 24.1 | 19.65 - 24.15 | 21.9 | 15 | 328.5 |
| 24.2 - 28.6 | 24.15 - 28.65 | 26.4 | 19 | 501.6 |
| 28.7 - 33.1 | 28.65 - 33.15 | 30.9 | 6 | 185.4 |
| 33.2 - 37.6 | 33.15 - 37.65 | 35.4 | 7 | 247.8 |
| 37.7 - 42.1 | 37.65 - 42.15 | 39.9 | 0 | 0 |
| 42.2 - 46.6 | 42.15 - 46.65 | 44.4 | 1 | 44.4 |
| | | | 51 | 1359.9 |

17. continued

a. $\overline{X} = \dfrac{\sum f \cdot X_m}{n} = \dfrac{1359.9}{51} = 26.66$

b. modal class: $24.2 - 28.6$

19.

| Class Limits | Boundaries | $X_m$ | $f$ | $f \cdot X_m$ |
|---|---|---|---|---|
| $13 - 19$ | $12.5 - 19.5$ | 16 | 2 | 32 |
| $20 - 26$ | $19.5 - 26.5$ | 23 | 7 | 161 |
| $27 - 33$ | $26.5 - 33.5$ | 30 | 12 | 360 |
| $34 - 40$ | $33.5 - 40.5$ | 37 | 5 | 185 |
| $41 - 47$ | $40.5 - 47.5$ | 44 | 6 | 264 |
| $48 - 54$ | $47.5 - 54.5$ | 51 | 1 | 51 |
| $55 - 61$ | $54.5 - 61.5$ | 58 | 0 | 0 |
| $62 - 68$ | $61.5 - 68.5$ | 65 | 2 | 130 |
| | | | 35 | 1183 |

a. $\overline{X} = \dfrac{\sum f \cdot X_m}{n} = \dfrac{1183}{35} = 33.8$

b. modal class: $27 - 33$

21.

| Boundaries | $X_m$ | $f$ | $f \cdot X_m$ |
|---|---|---|---|
| $15.5 - 18.5$ | 17 | 14 | 238 |
| $18.5 - 21.5$ | 20 | 12 | 240 |
| $21.5 - 24.5$ | 23 | 18 | 414 |
| $24.5 - 27.5$ | 26 | 10 | 260 |
| $27.5 - 30.5$ | 29 | 15 | 435 |
| $30.5 - 33.5$ | 32 | 6 | 192 |
| | | 75 | 1779 |

a. $\overline{X} = \dfrac{\sum f \cdot X_m}{n} = \dfrac{1779}{75} = 23.7$

b. modal class: $21.5 - 24.5$

23.

| Limits | Boundaries | $X_m$ | $f$ | $f \cdot X_m$ |
|---|---|---|---|---|
| 27 - 33 | 26.5 - 33.5 | 30 | 7 | 210 |
| 34 - 40 | 33.5 - 40.5 | 37 | 14 | 518 |
| 41 - 47 | 40.5 - 47.5 | 44 | 15 | 660 |
| 48 - 54 | 47.5 - 54.5 | 51 | 11 | 561 |
| 55 - 61 | 54.5 - 61.5 | 58 | 3 | 174 |
| 62 - 68 | 61.5 - 68.5 | 65 | 3 | 195 |
| 69 - 75 | 68.5 - 75.5 | 72 | 2 | 144 |
| | | | 55 | 2462 |

$\overline{X} = \dfrac{\sum f \cdot X_m}{n} = \dfrac{2462}{55} = 44.8$

modal class: $40.5 - 47.5$

25.

| Limits | Boundaries | $X_m$ | $f$ | $f \cdot X_m$ |
|--------|-----------|-------|-----|---------------|
| 1013 - 1345 | 1012.5 - 1345.5 | 1179 | 11 | 12969 |
| 1346 - 1678 | 1345.5 - 1678.5 | 1512 | 4 | 6048 |
| 1679 - 2011 | 1678.5 - 2011.5 | 1845 | 7 | 12915 |
| 2012 - 2344 | 2011.5 - 2344.5 | 2178 | 3 | 6534 |
| 2345 - 2677 | 2344.5 - 2677.5 | 2511 | 5 | 12555 |
| 2678 - 3010 | 2677.5 - 3010.5 | 2844 | 3 | 8532 |
| | | | 33 | 59553 |

$$\overline{X} = \frac{\sum f \cdot X_m}{n} = \frac{59553}{33} = 1804.6$$

modal class: $1013 - 1345$

27.
$$\overline{X} = \frac{\sum w \cdot X}{\sum w} = \frac{3(3.33) + 3(3.00) + 2(2.5) + 2.5(4.4) + 4(1.75)}{3 + 3 + 2 + 2.5 + 4} = \frac{41.99}{14.5} = 2.896$$

29.
$$\overline{X} = \frac{\sum w \cdot X}{\sum w} = \frac{9(427000) + 6(365000) + 12(725000)}{9 + 6 + 12} = \frac{14,733,000}{27} = \$545,666.67$$

31.
$$\overline{X} = \frac{\sum w \cdot X}{\sum w} = \frac{1(62) + 1(83) + 1(97) + 1(90) + 2(82)}{6} = \frac{496}{6} = 82.7$$

33.
a. Median     d. Mode
b. Mean       e. Mode
c. Mode      f. Mean

35.
Both could be true since one could be using the mean for the average salary, and the other could be using the mode for the average.

37.
$5 \cdot 8.2 = 41$
$6 + 10 + 7 + 12 + x = 41$
$x = 6$

39.
a. $\frac{2}{\frac{1}{30} + \frac{1}{45}} = 36$ mph

b. $\frac{2}{\frac{1}{40} + \frac{1}{25}} = 30.77$ mph

c. $\frac{2}{\frac{1}{50} + \frac{1}{10}} = \$16.67$

41.
$$\sqrt{\frac{8^2 + 6^2 + 3^2 + 5^2 + 4^2}{5}} = \sqrt{30} = 5.48$$

# Chapter 3 - Data Description

EXERCISE SET 3-2

**1.**
The square root of the variance is equal to the standard deviation.

**3.**
$\sigma^2, \sigma$

**5.**
When the sample size is less than 30, the formula for the true standard deviation of the sample will underestimate the population standard deviation.

**7.**
$R = 48 - 0 = 48$
$$s^2 = \frac{n\sum X^2 - (\sum X)^2}{n(n-1)} = \frac{10(4061) - 133^2}{10(10-1)} = \frac{22921}{90} = 254.68 \approx 254.7$$

$$s = \sqrt{254.7} = 15.96 \approx 16$$

**9.**
For Temperature:
$R = 61 - 29 = 32$
$$s^2 = \frac{n\sum X^2 - (\sum X)^2}{n(n-1)} = \frac{10(20{,}777) - 441^2}{10(10-1)} = 147.66$$

$$s = \sqrt{147.66} = 12.15$$

For Precipitation:
$R = 5.1 - 1.1 = 4.0$
$$s^2 = \frac{n\sum X^2 - (\sum X)^2}{n(n-1)} = \frac{10(86.13) - 26.3^2}{10(10-1)} = 1.88$$

$$s = \sqrt{1.88} = 1.37$$

Temperature is more variable.

**11.**
Houston:
$\overline{X} = \frac{725}{13} = 55.8$
$R = 75 - 47 = 28$
$$s^2 = \frac{n\sum X^2 - (\sum X)^2}{n(n-1)} = \frac{13(41{,}379) - 725^2}{13(13-1)} = \frac{12{,}302}{13(12)} = 78.859$$

$$s = \sqrt{78.859} = 8.88$$

Pittsburgh:
$\overline{X} = \frac{539}{13} = 41.5$
$R = 64 - 30 = 34$
$$s^2 = \frac{n\sum X^2 - (\sum X)^2}{n(n-1)} = \frac{13(23{,}413) - 539^2}{13(13-1)} = \frac{13{,}848}{13(12)} = 88.769$$

$$s = \sqrt{88.769} = 9.42$$

11. continued
The data for Pittsburgh seems more variable. Using the coefficient of variation to compare the data:
For Houston, $CVar = \frac{8.88}{55.8} = 15.91\%$. For Pittsburgh, $CVar = \frac{9.42}{41.5} = 22.7\%$
The data for Pittsburgh is more variable using the coefficients of variation.

13.
$R = 46 - 26 = 20$
Using the range rule of thumb, $s \approx \frac{20}{4} = 5$

15.
For 2004:
$R = 14{,}925 - 5044 = 9881$
$$s^2 = \frac{n\sum X^2 - (\sum X)^2}{n(n-1)} = \frac{24(1{,}917{,}585{,}062) - 202{,}108^2}{24(24-1)} = 9{,}373{,}909.1$$

$s = \sqrt{9{,}373{,}909.1} = 3061.7$

For 1990:
$R = 21{,}923 - 4768 = 17{,}155$
$$s^2 = \frac{n\sum X^2 - (\sum X)^2}{n(n-1)} = \frac{24(2{,}684{,}611{,}488) - 235{,}440^2}{24(24-1)} = 16{,}301{,}960.35$$

$s = \sqrt{16{,}301{,}960.35} = 4037.6$

The data for 1990 is more variable.

17.
$R = 11{,}413 - 150 = 11{,}263$
$$s^2 = \frac{n\sum X^2 - (\sum X)^2}{n(n-1)} = \frac{42(1{,}659{,}371{,}050) - 238{,}512^2}{42(42-1)} = 7{,}436{,}475.003$$

$s = \sqrt{7{,}436{,}475.003} = 2726.99$ or 2727

19.

| $X_m$ | f | $f \cdot X_m$ | $f \cdot X_m^2$ |
|---|---|---|---|
| 16 | 2 | 32 | 512 |
| 23 | 7 | 161 | 3703 |
| 30 | 12 | 360 | 10,800 |
| 37 | 5 | 185 | 6845 |
| 44 | 6 | 264 | 11,616 |
| 51 | 1 | 51 | 2601 |
| 58 | 0 | 0 | 0 |
| 65 | 2 | 130 | 8450 |
|  | 35 | 1183 | 44527 |

$$s^2 = \frac{n\sum f \cdot X_m^2 - (\sum f \cdot X_m)^2}{n(n-1)} = \frac{35(44{,}527) - 1183^2}{35(35-1)} = 133.58 \text{ or } 133.6$$

$s = \sqrt{133.58} = 11.6$

21.

| $X_m$ | f | $f \cdot X_m$ | $f \cdot X_m^2$ |
|---|---|---|---|
| 65 | 13 | 845 | 54,925 |
| 128 | 2 | 256 | 32,768 |
| 191 | 0 | 0 | 0 |
| 254 | 5 | 1270 | 322,580 |
| 317 | 1 | 317 | 100,489 |
| 380 | 1 | 380 | 144,400 |
| 443 | 0 | 0 | 0 |
| 506 | 1 | 506 | 256,036 |
| 569 | 2 | 1138 | 647,522 |
| | 25 | 4712 | 1,558,720 |

$$s^2 = \frac{n\sum f \cdot X_m^2 - (\sum f \cdot X_m)^2}{n(n-1)} = \frac{25(1,558,720) - 4712^2}{25(25-1)} = 27,941.76$$

$$s = \sqrt{27941.76} = 167.16 \text{ or } 167.2$$

23.

| $X_m$ | f | $f \cdot X_m$ | $f \cdot X_m^2$ |
|---|---|---|---|
| 58 | 6 | 348 | 20,184 |
| 69 | 12 | 828 | 57,132 |
| 80 | 25 | 2000 | 160,000 |
| 91 | 18 | 1638 | 148,058 |
| 102 | 14 | 1428 | 145,656 |
| 112 | 5 | 565 | 63,845 |
| | 80 | 6807 | 595,875 |

$$s^2 = \frac{n\sum f \cdot X_m^2 - (\sum f \cdot X_m)^2}{n(n-1)} = \frac{80(59,5875) - 6807^2}{80(80-1)} = 211.2$$

$$s = \sqrt{211.2} = 14.5$$

25.

| $X_m$ | f | $f \cdot X_m$ | $f \cdot X_m^2$ |
|---|---|---|---|
| 68 | 5 | 340 | 23,120 |
| 79 | 14 | 1106 | 87,374 |
| 90 | 18 | 1620 | 145,800 |
| 101 | 25 | 2525 | 255,025 |
| 112 | 12 | 1344 | 150,528 |
| 123 | 6 | 738 | 90,774 |
| | 80 | 7673 | 752,621 |

$$s^2 = \frac{n\sum f \cdot X_m^2 - (\sum f \cdot X_m)^2}{n(n-1)} = \frac{80(752,621) - 7673^2}{80(80-1)} = 211.19 \text{ or } 211.2$$

$$s = \sqrt{211.2} = 14.5$$

No, the variability of the lifetimes of the batteries is quite large.

27.

| $X_m$ | f | $f \cdot X_m$ | $f \cdot X_m^2$ |
|-------|---|---------------|-----------------|
| 27 | 5 | 135 | 3645 |
| 30 | 9 | 270 | 8100 |
| 33 | 32 | 1056 | 34848 |
| 36 | 30 | 720 | 25920 |
| 39 | 12 | 468 | 18252 |
| 62 | 2 | 84 | 3528 |
| | 80 | 2733 | 94293 |

$$s^2 = \frac{n\sum f \cdot X_m^2 - (\sum f \cdot X_m)^2}{n(n-1)} = \frac{80(94,293) - 2733^2}{80(80-1)} = 11.7$$

$$s = \sqrt{11.7} = 3.4$$

29.
For US: $\overline{X} = 3386.6$, s = 693.9; C. Var = $\frac{s}{\overline{X}} = \frac{693.9}{3386.6} = 0.2049$ or 20.49%

For World: $\overline{X} = 4997.8$, s = 803.2; C. Var = $\frac{s}{\overline{X}} = \frac{803.2}{4997.8} = 0.1607$ or 16.07%

The data for US is more variable.

31.
C. Var = $\frac{s}{\overline{X}} = \frac{6}{26} = 0.231 = 23.1\%$

C. Var = $\frac{s}{\overline{X}} = \frac{4000}{31,000} = 0.129 = 12.9\%$

Age is more variable.

33.
a. $1 - \frac{1}{5^2} = 0.96$ or 96%

b. $1 - \frac{1}{4^2} = 0.9375$ or 93.75%

35.
$\overline{X} = 240$ and s = 38
At least 75% of the data values will fall withing two standard deviations of the mean; hence,
$2(38) = 76$ and $240 - 76 = 164$ and $240 + 76 = 316$. Hence at least 75% of the
data values will fall between 164 and 316 calories.

37.
$1 - \frac{1}{k^2} = 0.8889$
$k = 3$
$\overline{X} = 640$ and s = 85
At least 88.89% of the data values will fall within 3 standard deviations of the mean, hence
$640 - 3(85) = 385$ and $640 + 3(85) = 895$. Therefore at least 88.89% of the data values will fall between
385 and 895 pounds.

39.
$\overline{X} = 12$ and s = 3
$20 - 12 = 8$ and $8 \div 3 = 2.67$
Hence, $1 - \frac{1}{k^2} = 1 - \frac{1}{2.67^2} = 1 - 0.14 = 0.86 = 86\%$
At least 86% of the data values will fall between 4 and 20.

**41.**
$26.8 + 1(4.2) = 31$
By the Empirical Rule, 68% of consumption is within 1 standard deviation of the mean. Then $\frac{1}{2}$ of 32%, or 16%, of consumption would be more than 31 pounds of citrus fruit per year.

**43.**
$n = 30$   $\overline{X} = 214.97$   $s = 20.76$   At least 75% of the data values will fall between $\overline{X} \pm 2s$.
$\overline{X} - 2(20.76) = 214.97 - 41.52 = 173.45$ and $\overline{X} + 2(20.76) = 214.97 + 41.52 = 256.49$
In this case all 30 values fall within this range; hence Chebyshev's Theorem is correct for this example.

**45.**
For $k = 1.5$, $1 - \frac{1}{1.5^2} = 1 - 0.44 = 0.56$ or 56%
For $k = 2$, $1 - \frac{1}{2^2} = 1 - 0.25 = 0.75$ or 75%
For $k = 2.5$, $1 - \frac{1}{2.5^2} = 1 - 0.16 = 0.84$ or 84%
For $k = 3$, $1 - \frac{1}{3^2} = 1 - 0.1111 = .8889$ or 88.89%
For $k = 3.5$, $1 - \frac{1}{3.5^2} = 1 - 0.08 = 0.92$ or 92%

**47.**
$\overline{X} = 13.3$
$$\text{Mean Dev} = \frac{|5-13.3|+|9-13.3|+|10-13.3|+|11-13.3|+|11-13.3|}{10}$$
$$+ \frac{|12-13.3|+|15-13.3|+|18-13.3|+|20-13.3|+|22-13.3|}{10} = 4.36$$

**49.**
For $n = 25$, $\overline{X} = 50$, and $s = 3$:

$$s\sqrt{n-1} = 3\sqrt{25-1} = 14.7 \qquad \overline{X} + s\sqrt{n-1} = 50 + 14.7 = 64.7$$

67 must be an incorrect data value, since is beyond the range using the formula $s\sqrt{n-1}$.

**EXERCISE SET 3-3**

**1.**
A z score tells how many standard deviations the data value is above or below the mean.

**3.**
A percentile is a relative measure while a percent is an absolute measure of the part to the total.

**5.**
$Q_1 = P_{25}$, $Q_2 = P_{50}$, $Q_3 = P_{75}$

**7.**
$D_1 = P_{10}$, $D_2 = P_{20}$, $D_3 = P_{30}$, etc

**9.**
For Canada:
$z = \frac{X-\overline{X}}{s} = \frac{26-29.4}{8.6} = -0.40$
For Italy:
$z = \frac{X-\overline{X}}{s} = \frac{42-29.4}{8.6} = 1.47$
For US:
$z = \frac{X-\overline{X}}{s} = \frac{13-29.4}{8.6} = -1.91$

11.

a. $z = \dfrac{X - \overline{X}}{s} = \dfrac{87 - 84}{4} = 0.75$

b. $z = \dfrac{79 - 84}{4} = -1.25$

c. $z = \dfrac{93 - 84}{4} = 2.25$

d. $z = \dfrac{76 - 84}{4} = -2$

e. $z = \dfrac{82 - 84}{4} = -0.5$

13.

For the statistics test:  $z = \dfrac{75 - 60}{10} = 1.5$

For the accounting test:  $z = \dfrac{36 - 30}{\sqrt{16}} = 1.5$

Neither.  The scores have the same relative position.

15.

a. $z = \dfrac{3.2 - 4.6}{1.5} = -0.93$    b. $z = \dfrac{630 - 800}{200} = -0.85$    c. $z = \dfrac{43 - 50}{5} = -1.4$

The score in part b is the highest.

17.

a. $24^{th}$    b. $67^{th}$    c. $48^{th}$    d. $88^{th}$

19.

a. 234    b. 251    c. 263    d. 274    e. 284

21.

a. $13^{th}$    b. $40^{th}$    c. $54^{th}$    d. $76^{th}$    e. $92^{nd}$

23.

$c = \dfrac{9(40)}{100} = 3.6$ or $4^{th}$ data value, which is 597

25.

$c = \dfrac{n \cdot p}{100} = \dfrac{7(60)}{100} = 4.2$ or 5    Hence, 47 is the closest value to the $60^{th}$ percentile.

27.

$c = \dfrac{10(40)}{100} = 4$        average the 4th and 5th values:  $P_{40} = \dfrac{2.1 + 2.2}{2} = 2.15$

29.

$c = \dfrac{6(33)}{100} = 1.98$ or $2^{nd}$ data value, which is 12.

31.

a.  5, 12, 16, 25, 32, 38        $Q_1 = 12$, $Q_2 = 20.5$, $Q_3 = 32$

Midquartile $= \dfrac{12 + 32}{2} = 22$        Interquartile range:  $32 - 12 = 20$

b.  53, 62, 78, 94, 96, 99, 103        $Q_1 = 62$, $Q_2 = 94$, $Q_3 = 99$

Midquartile $= \dfrac{62 + 99}{2} = 80.5$        Interquartile range:  $99 - 62 = 37$

EXERCISE SET 3-4

1. Data arranged in order: 6, 8, 12, 19, 27, 32, 54
Minimum: 6
$Q_1$: 8
Median: 19
$Q_3$: 32
Maximum: 54
Interquartile Range: $32 - 8 = 24$

3. Data arranged in order: 188, 192, 316, 362, 437, 589
Minimum: 188
$Q_1$: 192
Median: $\frac{316+362}{2} = 339$
$Q_3$: 437
Maximum: 589
Interquartile Range: $437 - 192 = 245$

5. Data arranged in order: 14.6, 15.5, 16.3, 18.2, 19.8
Minimum: 14.6
$Q_1$: $\frac{14.6+15.5}{2} = 15.05$
Median: 16.3
$Q_3$: $\frac{18.2+19.8}{2} = 19.0$
Maximum: 19.8
Interquartile Range: $19.0 - 15.05 = 3.95$

7. Minimum: 3
$Q_1$: 5
Median: 8
$Q_3$: 9
Maximum: 11
Interquartile Range: $9 - 5 = 4$

9. Minimum: 55
$Q_1$: 65
Median: 70
$Q_3$: 90
Maximum: 95
Interquartile Range: $90 - 65 = 25$

11.
MD $= \frac{30+31}{2} = 30.5$
$Q_1 = 29$   $Q_3 = 34$

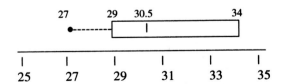

13.
MD $= \frac{253+289}{2} = 271$
$Q_1 = 238$        $Q_3 = 314$

13. continued

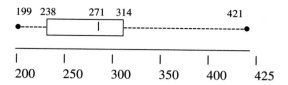

```
199  238      271  314                  421
●------┌──────┬─────┐--------------------------●
       └──────┼─────┘
┌─────────────────────────────────────
|       |       |       |       |       |
200    250     300     350     400     425
```

No, the distribution is not symmetric.

15.
$$MD = \frac{123 + 123}{2} = 123$$
$$Q_1 = 29.5 \qquad Q_3 = 135.5$$

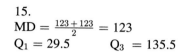

```
10  29.5            123  135.5                316
 ●-┌─────────────────┬─────┐-------------------------●
   └─────────────────┴─────┘
┌──────────────────────────────────────────
|       |       |       |       |       |       |       |
0       50     100     150     200     250     300     350
```

Based on the position of the median, the distribution is negatively skewed. Based on the lengths of the lines, the distribution is positively skewed.

17.
(a)
For April: $\overline{X} = 138$
For May: $\overline{X} = 391.7$
For June: $\overline{X} = 292$
For July: $\overline{X} = 143$
The month with the highest mean number of tornadoes is May.

(b)
For 2005: $\overline{X} = 177.3$
For 2004: $\overline{X} = 256.5$
For 2003: $\overline{X} = 289.8$
The year with the highest mean number of tornadoes is 2003.

(c) The 5-number summaries for each year are:
For 2005: 123, 127.5, 135, 227, 316
For 2004: 124, 124.5, 196.5, 388.5, 509
For 2003: 157, 162, 229.5, 417.5, 543

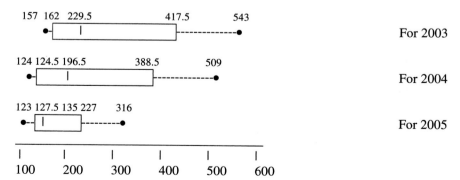

```
157 162  229.5              417.5        543
 ●-┌──┬───────────────────┐-------------●          For 2003

124 124.5 196.5       388.5        509
 ●-┌──┬───────────────┐-------------●              For 2004

123 127.5 135  227       316
●-┌┬──┐--------●                                   For 2005

┌────────────────────────────────────
|       |       |       |       |       |
100    200     300     400     500     600
```

17c. continued
The distribution for 2003, 2004, and 2005 are positively skewed. The data for 2005 appears to be the least variable.

REVIEW EXERCISES - CHAPTER 3

1.

a. $\overline{X} = \dfrac{\sum X}{n} = \dfrac{1273}{16} = 79.6$

b. 28, 35, 40, 53, 60, 62, 74, **78, 80**, 84, 84, 87, 108, 111, 123, 166

$MD = \dfrac{78+80}{2} = 79$

c. Mode = 84

d. $MR = \dfrac{28+166}{2} = 97$

e. Range = $166 - 28 = 138$

f. $s^2 = \dfrac{n\sum X^2 - (\sum X)^2}{n(n-1)} = \dfrac{16(120,173) - 1273^2}{16(16-1)} = 1259.3$

g. $s = \sqrt{1259.3} = 35.5$

3.

| Class | $X_m$ | f | $f \cdot X_m$ | $f \cdot X_m^2$ | cf |
|-------|-------|---|---------------|-----------------|----|
| 1 - 3 | 2 | 1 | 2 | 4 | 1 |
| 4 - 6 | 5 | 4 | 20 | 100 | 5 |
| 7 - 9 | 8 | 5 | 40 | 320 | 10 |
| 10 - 12 | 11 | 1 | 11 | 121 | 11 |
| 13 - 15 | 14 | 1 | 14 | 196 | 12 |
| | | 12 | 87 | 741 | |

a. $\overline{X} = \dfrac{\sum f \cdot X_m}{n} = \dfrac{87}{12} = 7.3$

b. Modal Class = $7 - 9$ or $6.5 - 9.5$

c. $s^2 = \dfrac{12(741) - 87^2}{12(11)} = \dfrac{1323}{132} = 10.0$

d. $s = \sqrt{10.0} = 3.2$

5.

| Class Boundaries | $X_m$ | f | $f \cdot X_m$ | $f \cdot X_m^2$ | cf |
|------------------|-------|---|---------------|-----------------|----|
| 12.5 - 27.5 | 20 | 6 | 120 | 2400 | 6 |
| 27.5 - 42.5 | 35 | 3 | 105 | 3675 | 9 |
| 42.5 - 57.5 | 50 | 5 | 250 | 12,500 | 14 |
| 57.5 - 72.5 | 65 | 8 | 520 | 33,800 | 22 |
| 72.5 - 87.5 | 80 | 6 | 480 | 38,400 | 28 |
| 87.5 - 102.5 | 95 | 2 | 190 | 18,050 | 30 |
| | | 30 | 1665 | 108,825 | |

a. $\overline{X} = \dfrac{\sum f \cdot X_m}{n} = \dfrac{1665}{30} = 55.5$

5. continued
b. Modal class $= 57.5 - 72.5$

c. $s^2 = \dfrac{n\sum f \cdot X_m^2 - (\sum f \cdot X_m)^2}{n(n-1)} = \dfrac{30(108{,}825) - 1665^2}{30(30-1)} = 566.1$

d. $s = \sqrt{566.1} = 23.8$

7.
$\overline{X} = \dfrac{\sum w \cdot X}{\sum w} = \dfrac{12 \cdot 0 + 8 \cdot 1 + 5 \cdot 2 + 5 \cdot 3}{12 + 8 + 5 + 5} = \dfrac{33}{30} = 1.1$

9.
$\overline{X} = \dfrac{\sum w \cdot X}{\sum w} = \dfrac{8 \cdot 3 + 1 \cdot 6 + 1 \cdot 30}{8 + 1 + 1} = \dfrac{60}{10} = 6$

11.
Magazines: C. Var $= \dfrac{s}{\overline{X}} = \dfrac{12}{56} = 0.214$ or $21.4\%$

Year: C. Var $= \dfrac{s}{\overline{X}} = \dfrac{2.5}{6} = 0.417$ or $41.7\%$

The number of years is more variable.

13.
a.

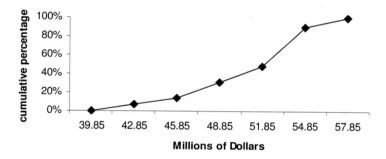

b. $P_{35} = 50$; $P_{65} = 53$; $P_{85} = 55$ (answers are approximate)
c. $44 = 10^{th}$ percentile; $48 = 26^{th}$ percentile; $54 = 78^{th}$ percentile
(answers are approximate)

15.
$\overline{X} = 0.32 \quad s = 0.03 \quad k = 2$
$0.32 - 2(0.03) = 0.26$ and $0.32 + 2(0.03) = 0.38$
At least 75% of the values will fall between \$0.26 and \$0.38.

17.
$\overline{X} = 54 \quad s = 4 \quad 60 - 54 = 6 \quad k = \dfrac{6}{4} = 1.5 \quad 1 - \dfrac{1}{1.5^2} = 1 - 0.44 = 0.56$ or $56\%$

19.
$\overline{X} = 32 \quad s = 4 \quad 44 - 32 = 12 \quad k = \dfrac{12}{4} = 3 \quad 1 - \dfrac{1}{3^2} = 0.8889 = 88.89\%$

21.
Top Nine Movies:
$MD = 2820 \quad Q_1 = 2520.5 \quad Q_3 = 3127.5$
Top Ten Movies:
$MD = 2699.5 \quad Q_1 = 2516 \quad Q_3 = 3044$

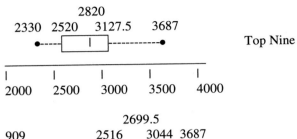

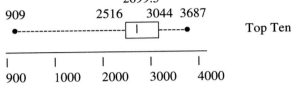

The range is much larger for the top ten movies.

23. By the Empirical Rule, 68% of the scores will be within 1 standard deviation of the mean.
$29.7 + 1(6) = 35.7$
$29.7 - 1(6) = 23.7$
Then 68% of commuters will get to work between 23.7 and 35.7 minutes.

CHAPTER 3 QUIZ
1. True
2. True
3. False
4. False
5. False
6. False
7. False
8. False
9. False
10. c.
11. c.
12. a. and b.
13. b.
14. d.
15. b.
16. statistic
17. parameters, statistics
18. standard deviation
19. $\sigma$
20. midrange
21. positively
22. outlier
23. a. 15.3    b. 15.5    c. 15, 16, 17    d. 15    e. 6    f. 3.61    g. 1.9
24. a. 6.4    b. $6-8$    c. 11.6    d. 3.4
25. a. 51.4    b. $35.5-50.5$    c. 451.5    d. 21.2
26. a. 8.2    b. $7-9$    c. 21.6    d. 4.6

27. 1.6
28. 4.46 or 4.5
29. 0.33; 0.162; newspapers
30. 0.3125; 0.229; brands
31. −0.75; −1.67; science
32. a. 0.5   b. 1.6   c. 15, c is higher
33. a. 56.25; 43.75; 81.25; 31.25; 93.75; 18.75; 6.25; 68.75   b. 0.9

c.

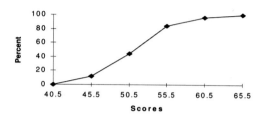

34.

a.

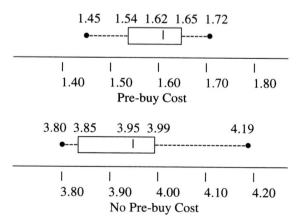

b. 47; 55; 64
c. $56^{th}$ percentile; $6^{th}$ percentile; $99^{th}$ percentile

35.
For Pre-buy:
MD = 1.62        $Q_1$ = 1.54        $Q_3$ = 1.65
For No Pre-buy:
MD = 3.95        $Q_1$ = 3.85        $Q_3$ = 3.99

```
      1.45   1.54 1.62 1.65 1.72
       •--------┌───┬──┐------•
                │   │  │
                └───┴──┘

   |      |      |      |      |
  1.40   1.50   1.60   1.70   1.80
            Pre-buy Cost
```

```
  3.80  3.85   3.95 3.99        4.19
    •---┌──┬────┐-------------------•
       │  │    │
       └──┴────┘

   |      |      |      |      |
  3.80   3.90   4.00   4.10   4.20
            No Pre-buy Cost
```

The cost of pre-buy gas is much less than to return the car without filling it with gas.  The variability of the return without filling with gas is larger than the variability of the pre-buy gas.

36.
For above 1129:  16%
For above 799:  97.5%

Note: Answers may vary due to rounding, TI-83's or computer programs.

EXERCISE SET 4-1

1.
A probability experiment is a chance process which leads to well-defined outcomes.

3.
An outcome is the result of a single trial of a probability experiment, whereas an event can consist of one or more outcomes.

5.
The range of values is $0 \leq P(E) \leq 1$.

7.
0

9.
$1 - 0.20 = 0.80$
Since the probability that it won't rain is 80%, you could leave your umbrella at home and be fairly safe.

11.
a. empirical          e. empirical
b. classical          f. empirical
c. empirical          g. subjective
d. classical

12.
a. $\frac{1}{6}$        d. 1
b. $\frac{1}{2}$        e. 1
c. $\frac{1}{3}$        f. $\frac{5}{6}$
g. $\frac{1}{6}$

13.
There are $6^2$ or 36 outcomes.
a. There are 5 ways to get a sum of 6. They are (1,5), (2,4), (3,3), (4,2), and (5,1). The probability then is $\frac{5}{36}$.

b. There are six ways to get doubles. They are (1,1), (2,2), (3,3), (4,4), (5,5), and (6,6). The probability then is $\frac{6}{36} = \frac{1}{6}$.

c. There are six ways to get a sum of 7. They are (1,6), (2,5), (3,4), (4,3), (5,2), and (6,1). There are two ways to get a sum of 11. They are (5,6) and (6,5). Hence, the total number of ways to get a 7 or 11 is eight. The probability then is $\frac{8}{36} = \frac{2}{9}$.

13. continued
d. To get a sum greater than nine, one must roll a 10, 11, or 12. There are six ways to get a 10, 11, or 12. They are (4,6), (5,5), (6,4), (5,6), (6,5), and (6,6). The probability then is $\frac{6}{36} = \frac{1}{6}$.

e. To get a sum less than or equal to four, one must roll a 4, 3, or 2. There are six ways to do this. They are (3,1), (2,2), (1,3), (2,1), (1,2), and (1,1). The probability is $\frac{6}{36} = \frac{1}{6}$.

14.
a. $\frac{1}{13}$          f. $\frac{4}{13}$

b. $\frac{1}{4}$          g. $\frac{1}{2}$

c. $\frac{1}{52}$          h. $\frac{1}{26}$

d. $\frac{2}{13}$          i. $\frac{7}{13}$

e. $\frac{4}{13}$          j. $\frac{1}{26}$

15.
There are 20 possible outcomes.

a. P(winning \$10) = P(rolling a 1)
P(rolling a 1) $= \frac{2}{20} = \frac{1}{10} = 0.1$

b. P(winning \$5 or \$10) = P(rolling either a 1 or 2)
P(1 or 1) $= \frac{4}{20} = \frac{1}{5} = 0.2$

c. P(winning a coupon) = P(rolling either a 3 or 4)
P(3 or 4) $= \frac{16}{20} = \frac{4}{5} = 0.8$

17.
a. P(type O) $= 0.43$

b. P(type A or B) $= 0.40 + 0.12 = 0.52$

c. P(not type A or O) $= 1 - 0.83 = 0.17$

19.
a. P(even prime number) $= \frac{1}{25} = 0.04$

b. P(sum of the digits is even) $= \frac{13}{25} = 0.52$

c. P(greater than 50) $= \frac{10}{25} = 0.4$

**21.**
The sample space is BBB, BBG, BGB, GBB, GGB, GBG, BGG, and GGG.

a. All boys is the outcome BBB; hence P(all boys) = $\frac{1}{8}$.

b. All girls or all boys would be BBB and GGG; hence, P(all girls or all boys) = $\frac{1}{4}$.

c. Exactly two boys or two girls would be BBG, BGB, GBB, BBG, GBG, or BGG. The probability then is $\frac{6}{8} = \frac{3}{4}$.

d. At least one child of each gender means at least one boy or at least one girl. The outcomes are the same as those of part c, hence the probability is the same, $\frac{3}{4}$.

**23.**
The outcomes for 2, 3, or 12 are (1,1), (1,2), (2,1), and (6,6); hence P(2, 3, or 12) = $\frac{1+2+1}{36} = \frac{4}{36} = \frac{1}{9}$.

**25.**
a. There are 18 odd numbers; hence, P(odd) = $\frac{18}{36} = \frac{9}{19}$.

b. There are 9 numbers greater than 27 (28 through 36) hence, the probability is $\frac{9}{38}$.

c. There are 5 numbers containing the digit 0 hence the probability is $\frac{5}{38}$.

d. The event in part a is most likely to occur since it has the highest probability of occurring.

**27.**
P(ten thousand dollar bill) = $\frac{3,460,000}{5,225,000} = 0.66$

**29.**
(a)

| | 1 | 2 | 3 | 4 | 5 | 6 |
|---|---|---|---|---|---|---|
| 1 | 1 | 2 | 3 | 4 | 5 | 6 |
| 2 | 2 | 4 | 6 | 8 | 10 | 12 |
| 3 | 3 | 6 | 9 | 12 | 15 | 18 |
| 4 | 4 | 8 | 12 | 16 | 20 | 24 |
| 5 | 5 | 10 | 15 | 20 | 25 | 30 |
| 6 | 6 | 12 | 18 | 24 | 30 | 36 |

**29. continued**
(b) P(multiple of 6) = $\frac{15}{36} = \frac{5}{12}$

(c) P(less than 10) = $\frac{17}{36}$

**31.**

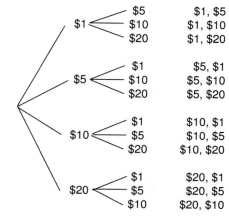

**33.**

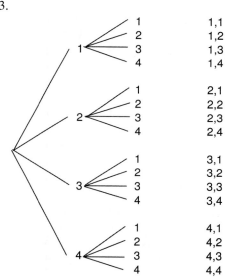

31

**35.**

English  Math  Elective

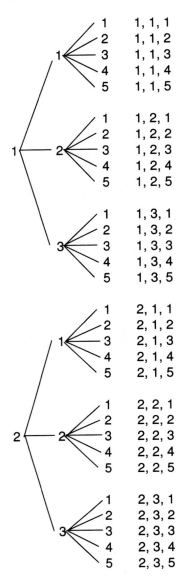

| | | |
|---|---|---|
| 1 | 1, 1, 1 |
| 2 | 1, 1, 2 |
| 3 | 1, 1, 3 |
| 4 | 1, 1, 4 |
| 5 | 1, 1, 5 |
| 1 | 1, 2, 1 |
| 2 | 1, 2, 2 |
| 3 | 1, 2, 3 |
| 4 | 1, 2, 4 |
| 5 | 1, 2, 5 |
| 1 | 1, 3, 1 |
| 2 | 1, 3, 2 |
| 3 | 1, 3, 3 |
| 4 | 1, 3, 4 |
| 5 | 1, 3, 5 |
| 1 | 2, 1, 1 |
| 2 | 2, 1, 2 |
| 3 | 2, 1, 3 |
| 4 | 2, 1, 4 |
| 5 | 2, 1, 5 |
| 1 | 2, 2, 1 |
| 2 | 2, 2, 2 |
| 3 | 2, 2, 3 |
| 4 | 2, 2, 4 |
| 5 | 2, 2, 5 |
| 1 | 2, 3, 1 |
| 2 | 2, 3, 2 |
| 3 | 2, 3, 3 |
| 4 | 2, 3, 4 |
| 5 | 2, 3, 5 |

**37.**
a. 0.08
b. 0.01
c. $0.08 + 0.27 = 0.35$
d. $0.01 + 0.24 + 0.11 = 0.36$

**39.**
The statement is probably not based on empirical probability and probably not true.

**41.**
Actual outcomes will vary, however each number should occur approximately $\frac{1}{6}$ of the time.

**43.**
a. 1:5, 5:1
b. 1:1, 1:1
c. 1:3, 3:1
d. 1:1, 1:1
e. 1:12, 12:1
f. 1:3, 3:1
g. 1:1, 1:1

**EXERCISE SET 4-2**

**1.**
Two events are mutually exclusive if they cannot occur at the same time. Examples will vary.

**3.**
a. $\frac{1,348,503}{1,907,172} = 0.707$

b. $\frac{46,024}{1,907,172} + \frac{1,098,371}{1,907,172} - \frac{21,683}{1,907,172} =$

$\frac{1,122,712}{1,907,172} = 0.589$

c. $\frac{21,683}{1,907,172} = 0.011$

d. $\frac{1,394,527}{1,907,172} = 0.731$

**5.**
$\frac{4}{19} + \frac{7}{19} = \frac{11}{19}$

**7.**
a. $\frac{56}{200} = \frac{7}{25}$ or 0.28
b. $\frac{75}{200} = \frac{3}{8}$ or 0.375
c. $\frac{34}{200} = \frac{17}{100}$ or 0.17
d. Event b has the highest probability so it is most likely to occur.

**9.**
P(football or basketball) =
$\frac{58 + 40 - 8}{200} = \frac{90}{200}$ or 0.45

P(neither) $= 1 - \frac{90}{200} = \frac{11}{20}$ or 0.55

**11.**

| | Junior | Senior | Total |
|---|---|---|---|
| Female | 6 | 6 | 12 |
| Male | 12 | 4 | 16 |
| Total | 18 | 10 | 28 |

a. $\frac{18}{28} + \frac{12}{28} - \frac{6}{28} = \frac{24}{28} = \frac{6}{7}$

**11. continued**

b. $\frac{10}{28} + \frac{12}{28} - \frac{6}{28} = \frac{16}{28} = \frac{4}{7}$

c. $\frac{18}{28} + \frac{10}{28} = \frac{28}{28} = 1$

**13.**

|        | 18 - 24 | 25 - 34 | Total  |
|--------|---------|---------|--------|
| Male   | 7922    | 2534    | 10,456 |
| Female | 5779    | 995     | 6,774  |
| Total  | 13,701  | 3529    | 17,230 |

a. P(female aged 25 - 34) $= \frac{995}{17,230} = 0.058$

b. P(male aged 18 - 24) $=$

$\frac{10,456}{17,230} + \frac{13,701}{17,230} - \frac{7922}{17,230} = \frac{16,235}{17,230} = 0.942$

c. P(under 25 years and not male) $=$

$\frac{5779}{17,230} = 0.335$

**15.**

Total $= 136,238$ multiple births
a. P(more than two babies) $=$
$\frac{7663}{136,328} = 0.056$

b. P(quads or quints) $= \frac{553}{136,328} = 0.004$

c. The total number of babies who are triplets $= 21,330$
The total number of babies from multiple births $= 280,957$
P(baby is a triplet) $= \frac{21,330}{280,957} = 0.076$

**17.**

|        | Ch. 6 | Ch. 8 | Ch. 10 | Total |
|--------|-------|-------|--------|-------|
| Quiz   | 5     | 2     | 1      | 8     |
| Comedy | 3     | 2     | 8      | 13    |
| Drama  | 4     | 4     | 2      | 10    |
| Total  | 12    | 8     | 11     | 31    |

a. P(quiz show or channel 8) = P(quiz) + P(channel 8) − P(quiz show on ch. 8) $=$
$\frac{8}{31} + \frac{8}{31} - \frac{2}{31} = \frac{14}{31}$

b. P(drama or comedy) = P(drama) + P(comedy) $= \frac{13}{31} + \frac{10}{31} = \frac{23}{31}$

c. P(channel 10 or drama) = P(ch. 10) + P(drama) − P(drama on channel 10) $=$
$\frac{11}{31} + \frac{10}{31} - \frac{2}{31} = \frac{19}{31}$

**19.**

The total of the frequencies is 30.
a. $\frac{2}{30} = \frac{1}{15}$

b. $\frac{2+3+5}{30} = \frac{10}{30} = \frac{1}{3}$

c. $\frac{12+8+2+3}{30} = \frac{25}{30} = \frac{5}{6}$

d. $\frac{12+8+2+3}{30} = \frac{25}{30} = \frac{5}{6}$

e. $\frac{8+2}{30} = \frac{10}{30} = \frac{1}{3}$

**21.**

The total of the frequencies is 24.
a. $\frac{10}{24} = \frac{5}{12}$

b. $\frac{2+1}{24} = \frac{3}{24} = \frac{1}{8}$

c. $\frac{10+3+2+1}{24} = \frac{16}{24} = \frac{2}{3}$

d. $\frac{8+10+3+2}{24} = \frac{23}{24}$

**23.**

a. There are 4 kings, 4 queens, and 4 jacks; hence P(king or queen or jack) $= \frac{12}{52} = \frac{3}{13}$

b. There are 13 clubs, 13 hearts, and 13 spades; hence, P(club or heart or spade) $=$
$\frac{13+13+13}{52} = \frac{39}{52} = \frac{3}{4}$

c. There are 4 kings, 4 queens, and 13 diamonds but the king and queen of diamonds were counted twice, hence; P(king or queen or diamond) = P(king) + P(queen) + P(diamond) − P(king and queen of diamonds) $= \frac{4}{52} + \frac{4}{52} + \frac{13}{52} - \frac{2}{52} = \frac{19}{52}$

d. There are 4 aces, 13 diamonds, and 13 hearts. There is one ace of diamonds and one ace of hearts; hence, P(ace or diamond or heart) = P(ace) + P(diamond) + P(heart) − P(ace of hearts and ace of diamonds) $=$
$\frac{4}{52} + \frac{13}{52} + \frac{13}{52} - \frac{2}{52} = \frac{28}{52} = \frac{7}{13}$

e. There are 4 nines, 4 tens, 13 spades, and 13 clubs. There is one nine of spades, one ten of spades, one nine of clubs and one ten of clubs. Hence, P(9 or 10 or spade or club) = P(9) + P(10) + P(spade) + P(club) − P(9 and 10 of clubs and spades) $=$
$\frac{4}{52} + \frac{4}{52} + \frac{13}{52} + \frac{13}{52} - \frac{4}{52} = \frac{30}{52} = \frac{15}{26}$

**25.**
P(export or
ethanol) $= \frac{1.9}{11} + \frac{1.6}{11} - \frac{0}{11} = 0.318$

**27.**
P(mushrooms or pepperoni) $=$
$\quad$ P(mushrooms) $+$ P(pepperoni) $-$
$\quad$ P(mushrooms and pepperoni)

Let X = P(mushrooms and pepperoni)
Then $0.55 = 0.32 + 0.17 - X$
$X = 0.06$

**29.**
P(not a two-car garage) $= 1 - 0.70 = 0.30$

**EXERCISE SET 4-3**

**1.**
a. independent$\quad$e. independent
b. dependent$\quad\quad$f. dependent
c. dependent$\quad\quad$g. dependent
d. dependent$\quad\quad$h. independent

**3.**
a. P(none play video or computer
games) $= (0.31)^4 = 0.009$ or $0.9\%$

b. P(all four play video or computer
games) $= (0.69)^4 = 0.227$ or $22.7\%$

**5.**
P(male graduate) $= 1 - 0.28 = 0.72$
P(all 3 are male) $= (0.72)^3 = 0.373$ or
$37.3\%$
The event is unlikely to occur since the
probability is less than 0.5.

**7.**
a. P(no computer) $= 1 - 0.543 = 0.457$
P(none of three has a computer) $=$
$(0.457)^3 = 0.0954$

b. P(at least one has a computer) $=$
$1 -$ P(none of three has a computer) $=$
$1 - 0.0954 = 0.9046$

c. P(all three have computers) $=$
$(0.543)^3 = 0.1601$

**9.**
P(all are citizens) $= (0.801)^3 = 0.5139$

**11.**
a. P(at least one doesn't use a computer at
work) $= 1 -$ P(all of the women use a
computer at work)
P(at least one doesn't use a
computer) $= 1 - (0.72)^5 = 0.807$

b. P(all 5 use a
computer) $= (0.72)^5 = 0.1935$

**13.**
a. P(all 3 get enough
exercise) $= (0.27)^3 = 0.0197$

b. P(at least one gets enough
exercise) $= 1 - (0.73)^3 = 0.611$

**15.**
P(5 buy at least 1) $= (\frac{90}{120})^5 = \frac{243}{1024}$

**17.**
$\frac{5}{8} \cdot \frac{4}{7} \cdot \frac{3}{6} = \frac{5}{28}$

**19.**
$\frac{23}{38} \cdot \frac{22}{37} \cdot \frac{21}{36} = \frac{1771}{8436}$ or $0.210$
The event is unlikely to occur since the
probability is less than 0.5.

**21.**

P(defective) $= 0.08 + 0.036 = 0.116$

23.

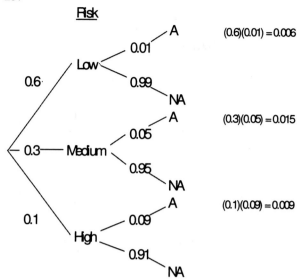

<u>Risk</u>

(0.6)(0.01) = 0.006

(0.3)(0.05) = 0.015

(0.1)(0.09) = 0.009

P(accident) = .006 + .015 + .009 = 0.03

25.
P(red ball) = $\frac{1}{3} \cdot \frac{5}{8} + \frac{1}{3} \cdot \frac{3}{4} + \frac{1}{3} \cdot \frac{4}{6} = \frac{49}{72}$

27.
P( ≤ 9 1st roll and ≤ 9 2nd roll and >9 3rd roll) = $\frac{30}{36} \cdot \frac{30}{36} \cdot \frac{6}{36} = 0.116$

29.
P(swim | bridge) = $\frac{\text{P(play bridge and swim)}}{\text{P(play bridge)}}$

= $\frac{0.73}{0.82} = 0.89$ or 89%

31.
P(garage | deck) = $\frac{0.42}{0.60} = 0.7$ or 70%

33.
a. P(coffee or candy) = $\frac{43}{77} + \frac{22}{77} - \frac{10}{77} = 0.714$

b. P(tea | contains mugs) = $\frac{10/77}{23/77} = 0.435$

c. P(tea and cookies) = $\frac{12}{77} = 0.156$

35.
a. P(male | pediatrician) = $\frac{\frac{33,020}{124,645}}{\frac{66,371}{124,645}} = 0.498$

b. P(pathologist | female) = $\frac{5604}{51,247} = 0.109$

c. No. P(pathologist | female) ≠ P(female)

37.
a. P(none have been married) = $(0.703)^5 = 0.1717$

b. P(at least one has been married) =
1 − P(none have been married)
= 1 − 0.1717
= 0.8283

39.
P(at least one not on time) =
1 − P(none not on time)
= 1 − P(all 5 on time)
= 1 − ().843)^5 = 0.574

41.
If P(read to) = 0.58, then
P(not being read to) = 1 − 0.58 = 0.42

P(at least one is read to) = 1 − P(none are read to)
= 1 − P(all five are not read to)
= 1 − (0.42)^5 = 0.9869

43.
P(at least one club) = 1 − P(no clubs)
$1 - \frac{39}{52} \cdot \frac{38}{51} \cdot \frac{37}{50} \cdot \frac{36}{49} = 1 - \frac{6327}{20,825}$
$= \frac{14,498}{20,825}$

45.
a. P(not a family and children's game) = 1 − 0.198 = 0.802
P(none of five are family and children's games) = $(0.802)^5 = 0.332$

b. P(at least one is family and children's game) = 1 − 0.332 = 0.668

47.
P(at least one tail) = 1 − P(no tails)
$1 - (\frac{1}{2})^5 = 1 - \frac{1}{32} = \frac{31}{32}$

49.
P(at least one 3) = 1 − P(no 3's)
$1 - (\frac{5}{6})^7 = 1 - \frac{78,125}{279,936} = \frac{201,811}{279,936}$ or 0.721
The event is likely to occur since the probability is about 72%.

51.
P(at least one even) = 1 − P(no evens)
$1 - (\frac{1}{2})^3 = 1 - \frac{1}{8} = \frac{7}{8}$

**53.**
No, because $P(A \cap B) = 0$ therefore
$P(A \cap B) \neq P(A) \cdot P(B)$

**55.**
Yes.

$P(\text{enroll}) = 0.55$

$P(\text{enroll} \mid DW) > P(\text{enroll})$ which indicates that DW has a positive effect on enrollment.

$P(\text{enroll} \mid LP) = P(\text{enroll})$ which indicates that LP has no effect on enrollment.

$P(\text{enroll} \mid MH) < P(\text{enroll})$ which indicates that MH has a low effect on enrollment.

Thus, all students should meet with DW.

**EXERCISE SET 4-4**

**1.**
$10^5 = 100,000$
$10 \cdot 9 \cdot 8 \cdot 7 \cdot 6 = 30,240$

**3.**
$7! = 7 \cdot 6 \cdot 5 \cdot 4 \cdot 3 \cdot 2 \cdot 1 = 5040$

**5.**
$8! = 8 \cdot 7 \cdot 6 \cdot 5 \cdot 4 \cdot 3 \cdot 2 \cdot 1 = 40,320$

**7.**
$7! = 7 \cdot 6 \cdot 5 \cdot 4 \cdot 3 \cdot 2 \cdot 1 = 5040$

**9.**
$10 \cdot 10 \cdot 10 = 1000$
$1 \cdot 9 \cdot 8 = 72$

**11.**
$6 \cdot 5 \cdot 5 \cdot 4 = 600$

**13.**
a. $8! = 8 \cdot 7 \cdot 6 \cdot 5 \cdot 4 \cdot 3 \cdot 2 \cdot 1 = 40,320$

b. $10! = 10 \cdot 9 \cdot 8 \cdot 7 \cdot 6 \cdot 5 \cdot 4 \cdot 3 \cdot 2 \cdot 1$
$10! = 3,628,800$

c. $0! = 1$

d. $1! = 1$

e. $_7P_5 = \frac{7!}{(7-5)!}$

**13e. continued**

$_7P_5 = \frac{7 \cdot 6 \cdot 5 \cdot 4 \cdot 3 \cdot 2 \cdot 1}{2 \cdot 1} = 2520$

f. $_{12}P_4 = \frac{12!}{(12-4)!}$

$= \frac{12 \cdot 11 \cdot 10 \cdot 9 \cdot 8 \cdot 7 \cdot 6 \cdot 5 \cdot 4 \cdot 3 \cdot 2 \cdot 1}{8 \cdot 7 \cdot 6 \cdot 5 \cdot 4 \cdot 3 \cdot 2 \cdot 1} = 11,880$

g. $_5P_3 = \frac{5!}{(5-3)!}$

$= \frac{5 \cdot 4 \cdot 3 \cdot 2 \cdot 1}{2 \cdot 1} = 60$

h. $_6P_0 = \frac{6!}{(6-0)!}$

$= \frac{6 \cdot 5 \cdot 4 \cdot 3 \cdot 2 \cdot 1}{6 \cdot 5 \cdot 4 \cdot 3 \cdot 2 \cdot 1} = 1$

i. $_5P_5 = \frac{5!}{(5-5)!}$

$= \frac{5 \cdot 4 \cdot 3 \cdot 2 \cdot 1}{0!} = 120$

j. $_6P_2 = \frac{6!}{(6-2)!}$

$= \frac{6 \cdot 5 \cdot 4 \cdot 3 \cdot 2 \cdot 1}{4 \cdot 3 \cdot 2 \cdot 1} = 30$

**15.**
$_4P_4 = \frac{4!}{(4-4)!} = \frac{4 \cdot 3 \cdot 2 \cdot 1}{0!} = 24$

**17.**
$_{22}C_4 = \frac{22!}{(22-4)!\,4!} = 7315$

**19.**
$_7P_4 = \frac{7!}{(7-4)!} = \frac{7 \cdot 6 \cdot 5 \cdot 4 \cdot 3 \cdot 2 \cdot 1}{3 \cdot 2 \cdot 1} = 840$

**21.**
$_{10}P_6 = \frac{10!}{(10-6)!} = \frac{10 \cdot 9 \cdot 8 \cdot 7 \cdot 6 \cdot 5 \cdot 4 \cdot 3 \cdot 2 \cdot 1}{4 \cdot 3 \cdot 2 \cdot 1} = 151,200$

**23.**
$_{50}P_4 = \frac{50!}{(50-4)!} = \frac{50!}{46!} = 5,527,200$

**25.**
Same task: $_{12}C_4 = \frac{12!}{(12-4)!\,4!} = 495$

Different tasks: $_{12}P_4 = \frac{12!}{(12-4)!} = 11,880$

**27.**
a. $\frac{5!}{3!\,2!} = 10$     f. $\frac{3!}{3!\,0!} = 1$

b. $\frac{8!}{5!\,3!} = 56$     g. $\frac{3!}{0!\,3!} = 1$

c. $\frac{7!}{3!\,4!} = 35$     h. $\frac{9!}{2!\,7!} = 36$

**27. continued**

d. $\frac{6!}{4!\,2!} = 15$   i. $\frac{12!}{10!\,2!} = 66$

e. $\frac{6!}{2!\,4!} = 15$   j. $\frac{4!}{1!\,3!} = 4$

**29.**

$_{10}C_3 = \frac{10!}{7!\,3!} = \frac{10\cdot9\cdot8\cdot7!}{7!\cdot3\cdot2\cdot1} = 120$

**31.**

$_{10}C_4 = \frac{10!}{6!\,4!} = 210$

**33.**

$_{20}C_5 = \frac{20!}{15!\,5!} = 15{,}504$

**35.**

Different programs:
$_{18}C_{10} = \frac{18!}{(18-10)!\,10!} = 43{,}758$

Starting and ending with the same song:
$_{16}C_8 = \frac{16!}{(16-8)!\,8!} = 12{,}870$

**37.**

$_{12}C_4 = 495$
$_7C_2 \cdot {_5}C_2 = 21 \cdot 10 = 210$
$_7C_2 \cdot {_5}C_2 + {_7}C_3 \cdot {_5}C_1 + {_7}C_4 =$
$21 \cdot 10 + 35 \cdot 5 + 35 =$
$210 + 175 + 35 = 420$

**39.**

The possibilities are CVV or VCV or VVV.

Assuming the same vowel can't be used
twice in a "word":
$7 \cdot 5 \cdot 4 + 5 \cdot 7 \cdot 4 + 5 \cdot 4 \cdot 3 = 340$

Assuming the same vowel can be used twice
in a "word":
$7 \cdot 5 \cdot 5 + 5 \cdot 7 \cdot 5 + 5 \cdot 5 \cdot 5 = 475$

**41.**

$_{10}C_2 \cdot {_{12}}C_2 = \frac{10!}{8!\,2!} \cdot \frac{12!}{10!\,2!}$
$= 45 \cdot 66 = 2{,}970$

**43.**

There are $_7C_2 = 21$ tiles with unequal
numbers and 7 tiles with equal numbers.
Thus, the total number of tiles is 28.

**45.**

$_{11}C_7 = \frac{11!}{4!\,7!} = \frac{11\cdot10\cdot9\cdot8\cdot7!}{7!\cdot4\cdot3\cdot2\cdot1} = 330$

**47.**

$_{10}C_6 \cdot {_{12}}C_6 = \frac{10!}{4!\,6!} \cdot \frac{12!}{6!\,6!} = 194{,}040$

**49.**

$_{20}C_8 = \frac{20!}{(20-8)!\,8!}$
$= \frac{20\cdot19\cdot18\cdot17\cdot16\cdot15\cdot14\cdot13\cdot12!}{12!\cdot8\cdot7\cdot6\cdot5\cdot4\cdot3\cdot2\cdot1} = 125{,}970$

**51.**

$_{20}P_5 = \frac{20!}{15!} = \frac{20\cdot19\cdot18\cdot17\cdot16\cdot15!}{15!} = 1{,}860{,}480$

**53.**

$_{17}C_2 = \frac{17!}{(17-2)!\,2!} = \frac{17\cdot16\cdot15!}{15!\cdot2\cdot1} = 136$

**55.**

$_5P_5 = \frac{5!}{0!} = \frac{5\cdot4\cdot3\cdot2\cdot1}{1} = 120$

**57.**

$_6C_3 \cdot {_5}C_2 = \frac{6!}{3!\,3!} \cdot \frac{5!}{3!\,2!} = 200$

**59.**

$_8P_3 = \frac{8!}{5!} = \frac{8\cdot7\cdot6\cdot5\cdot4\cdot3\cdot2\cdot1}{5\cdot4\cdot3\cdot2\cdot1} = 336$

**EXERCISE SET 4-5**

**1.**
$P(\text{2 face cards}) = \frac{12}{52} \cdot \frac{11}{51} = \frac{11}{221}$

**3.**

a. There are $_4C_3$ ways of selecting 3 women
and $_7C_3$ total ways to select 3 people;
hence, $P(\text{all women}) = \frac{_4C_3}{_7C_3} = \frac{4}{35}$.

b. There are $_3C_3$ ways of selecting 3 men;
hence, $P(\text{all men}) = \frac{_3C_3}{_7C_3} = \frac{1}{35}$.

c. There are $_3C_2$ ways of selecting 2 men
and $_4C_1$ ways of selecting one woman;
hence, $P(\text{2 men and 1 woman}) = \frac{_3C_2 \cdot {_4}C_1}{_7C_3}$
$= \frac{12}{35}$.

d. There are $_3C_1$ ways to select one man and
$_4C_2$ ways of selecting two women; hence,
$P(\text{1 man and 2 women}) = \frac{_3C_1 \cdot {_4}C_2}{_7C_3} = \frac{18}{35}$.

**5.**

a. $P(\text{all Republicans}) =$
$\frac{_{51}C_3 \cdot {_{48}}C_0 \cdot {_1}C_0}{_{100}C_3} = \frac{20{,}825}{161{,}700} = 0.129$

b. $P(\text{all Democrats}) =$
$\frac{_{51}C_0 \cdot {_{48}}C_3 \cdot {_1}C_0}{_{100}C_3} = \frac{17{,}296}{161{,}700} = 0.107$

**5. continued**

c. P(one Republican, Democrat, and Independent) =

$\frac{_{51}C_1 \cdot _{48}C_1 \cdot _1C_1}{_{100}C_3} = \frac{2448}{161,700} = 0.015$

**7.**

$\frac{2}{50} \cdot \frac{1}{49} = \frac{1}{1225}$

**9.**

a. P(one of each) =

$\frac{_{18}C_1 \cdot _{10}C_1 \cdot _3C_1}{_{31}C_3} = \frac{540}{4495} = 0.120$

b. P(no Navy members) =

$\frac{_{21}C_3}{_{31}C_3} = \frac{1330}{4495} = 0.296$

c. P(three Marines) =

$\frac{_{18}C_3 \cdot _{10}C_0 \cdot _3C_0}{_{31}C_3} = \frac{816}{4495} = 0.182$

**11.**

a. P(red) $= \frac{_{11}C_2}{_{19}C_2} = \frac{55}{171} = 0.3216$

b. P(black) $= \frac{_8C_2}{_{19}C_2} = \frac{28}{171} = 0.1637$

c. P(unmatched) $= \frac{_{11}C_1 \cdot _8C_1}{_{19}C_2} = \frac{88}{171} = 0.5146$

d. It probably got lost in the wash!

**13.**

There are $6^3 = 216$ ways of tossing three dice, and there are 15 ways of getting a sum of 7; i.e., (1, 1, 5), (1, 5, 1), (5, 1, 1), (1, 2, 4), etc. Hence the probability of rolling a sum of 7 is $\frac{15}{216} = \frac{5}{72}$.

**15.**

There are $5! = 120$ ways to arrange 5 washers in a row and 2 ways to have them in correct order, small to large or large to small; hence, the probability is $\frac{2}{120} = \frac{1}{60}$.

**17.**

P(berries are produced) = P(either 1 or 2 males)

P(1 or 2 males) $= \frac{_8C_2 \cdot _4C_1}{_{12}C_3} + \frac{_8C_1 \cdot _4C_2}{_{12}C_3}$

$= 0.509 + 0.218 = 0.727$

**REVIEW EXERCISES - CHAPTER 4**

**1.**

a. $\frac{1}{6}$    b. $\frac{1}{6}$    c. $\frac{3}{6} = \frac{1}{2}$

**3.**

a. P(not used for taxes) = P(virus or other)

P(virus or other) $= \frac{5}{10} + \frac{2}{10} = 0.7$

b. P(taxes or other use) $= \frac{3}{10} + \frac{2}{10} = 0.5$

**5.**

$\frac{850}{1500} = \frac{17}{30}$

**7.**

P(either air-conditioning or CD player)

$= 0.5 + 0.37 - 0.06 = 0.81$

P(neither air-conditioning nor CD)

$= 1 - 0.81 = 0.19$

**9.**

$0.80 + 0.30 - 0.12 = 0.98$

**11.**

P(enrolled in an online course) $= \frac{1}{6}$ or 0.167

a. P(all 5 took an online course) $= (\frac{1}{6})^5 = 0.0001$

b. P(none took an online course) $= (\frac{5}{6})^5 = 0.402$

c. P(at least one took an online course)

$= 1 - $ P(none took an online course)

$= 1 - (\frac{5}{6})^5 = 0.598$

**13.**

a. $\frac{26}{52} \cdot \frac{25}{51} \cdot \frac{24}{50} = \frac{2}{17}$

b. $\frac{13}{52} \cdot \frac{12}{51} \cdot \frac{11}{50} = \frac{33}{2550} = \frac{11}{850}$

c. $\frac{4}{52} \cdot \frac{3}{51} \cdot \frac{2}{50} = \frac{1}{5525}$

**15.**

Total number of movie releases $= 1384$

a. P(European) $= \frac{834}{1384} = 0.603$

b. P(US) $= \frac{471}{1384} = 0.340$

c. P(German or French) $= \frac{316}{1384} + \frac{132}{1384}$

$= \frac{448}{1384}$ or $0.324$

d. P(German | European)

$= \frac{\text{P(European and German)}}{\text{P(European)}} = \frac{\frac{316}{1384}}{\frac{834}{1384}} = 0.379$

17.

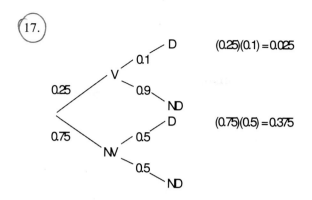

$(0.25)(0.1) = 0.025$

$(0.75)(0.5) = 0.375$

$P(\text{disease}) = 0.025 + 0.375 = 0.4$

19.
$P(NC \mid C) = \frac{P(NC \text{ and } C)}{P(C)} = \frac{0.37}{0.73} = 0.51$

21.
$\frac{0.43}{0.75} = 0.573$ or $57.3\%$

23.

|            | <4 yrs HS | HS | College | Total |
|------------|-----------|-----|---------|-------|
| Smoker     | 6         | 14  | 19      | 39    |
| Non-Smoker | 18        | 7   | 25      | 50    |
| Total      | 24        | 21  | 44      | 89    |

a. There are 44 college graduates and 19 of them smoke; hence, the probability is $\frac{19}{44}$.

b. There are 24 people who did not graduate from high school, 6 of whom do not smoke; hence, the probability is
$\frac{6}{24} = \frac{1}{4}$.

25.
P(at least one household has no DVD player)
$= 1 - P(\text{none have no DVD player})$
$= 1 - P(\text{all 6 have DVD players})$
$= 1 - (0.81)^6 = 0.718$

27.
If repetitions are allowed:
$26 \cdot 26 \cdot 26 \cdot 10 \cdot 10 \cdot 10 = 175,760,000$

If repetitions are not allowed:
$_{26}P_3 \cdot {_{10}}P_4 = \frac{26 \cdot 25 \cdot 24 \cdot 23!}{23!} \cdot \frac{10 \cdot 9 \cdot 8 \cdot 7 \cdot 6!}{6!}$
$= 78,624,000$

If repetitions are allowed in the letters but not in the digits:
$26 \cdot 26 \cdot 26 \cdot {_{10}}P_4 = 88,583,040$

29.
$_{5}C_3 \cdot {_{7}}C_4 = \frac{5!}{2! \, 3!} \cdot \frac{7!}{3! \, 4!} = 10 \cdot 35 = 350$

31.
$_{10}C_2 = \frac{10!}{8! \, 2!} = 45$

33.
$100!$

35.
$_{12}C_4 = \frac{12!}{8! \, 4!} = \frac{12 \cdot 11 \cdot 10 \cdot 9 \cdot 8!}{4 \cdot 3 \cdot 2 \cdot 1 \cdot 8!} = 495$

37.
$_{20}C_5 = \frac{20!}{15! \, 5!} = \frac{20 \cdot 19 \cdot 18 \cdot 17 \cdot 16 \cdot 15!}{15! \, 5 \cdot 4 \cdot 3 \cdot 2 \cdot 1} = 15,504$

39.
Total number of outcomes:
$26 \cdot 26 \cdot 26 \cdot 10 \cdot 10 \cdot 10 \cdot 10 = 175,760,000$

Total number of ways for USA followed by a number divisible by 5:
$1 \cdot 1 \cdot 1 \cdot 10 \cdot 10 \cdot 10 \cdot 2 = 2000$

Hence $P = \frac{2000}{175,760,000} = 0.0000114$

41.
Total number of territories $= 45$

P(3 French or 3 UK or 3 US) $= \frac{{_{16}}C_3}{{_{45}}C_3} + \frac{{_{15}}C_3}{{_{45}}C_3} + \frac{{_{14}}C_3}{{_{45}}C_3}$

$= \frac{560}{14,190} + \frac{455}{14,190} + \frac{364}{14,190}$

$= \frac{1379}{14,190} = 0.097$

43.

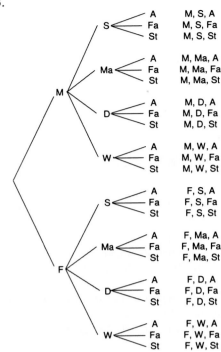

| | | | |
|---|---|---|---|
| | S | A | M, S, A |
| | | Fa | M, S, Fa |
| | | St | M, S, St |
| | Ma | A | M, Ma, A |
| | | Fa | M, Ma, Fa |
| | | St | M, Ma, St |
| M | D | A | M, D, A |
| | | Fa | M, D, Fa |
| | | St | M, D, St |
| | W | A | M, W, A |
| | | Fa | M, W, Fa |
| | | St | M, W, St |
| | S | A | F, S, A |
| | | Fa | F, S, Fa |
| | | St | F, S, St |
| | Ma | A | F, Ma, A |
| | | Fa | F, Ma, Fa |
| | | St | F, Ma, St |
| F | D | A | F, D, A |
| | | Fa | F, D, Fa |
| | | St | F, D, St |
| | W | A | F, W, A |
| | | Fa | F, W, Fa |
| | | St | F, W, St |

## CHAPTER 4 QUIZ

1. False, subjective probability can be used when other types of probabilities cannot be found.
2. False, empirical probability uses frequency distributions.
3. True
4. False, $P(A \text{ or } B) = P(A) + P(B) - P(A \text{ and } B)$
5. False, the probabilities can be different.
6. False, complementary events cannot occur at the same time.
7. True
8. False, order does not matter in combinations.
9. b.
10. b. and d.
11. d.
12. b.
13. c.
14. b.
15. d.
16. b.
17. b.
18. sample space
19. zero and one
20. zero
21. one
22. mutually exclusive

23. a. $\frac{4}{52} = \frac{1}{13}$   c. $\frac{16}{52} = \frac{4}{13}$

   b. $\frac{4}{52} = \frac{1}{13}$

24. a. $\frac{13}{52} = \frac{1}{4}$   d. $\frac{4}{52} = \frac{1}{13}$

   b. $\frac{4+13-1}{52} = \frac{4}{13}$   e. $\frac{26}{52} = \frac{1}{2}$

   c. $\frac{1}{52}$

25. a. $\frac{12}{31}$   c. $\frac{27}{31}$

   b. $\frac{12}{31}$   d. $\frac{24}{31}$

26. a. $\frac{11}{36}$   d. $\frac{1}{3}$

   b. $\frac{5}{18}$   e. 0

   c. $\frac{11}{36}$   f. $\frac{11}{12}$

27. $(0.75 - 0.16) + (0.25 - 0.16) = 0.68$

28. $(0.3)^5 = 0.002$

29. a. $\frac{26}{52} \cdot \frac{25}{51} \cdot \frac{24}{50} \cdot \frac{23}{49} \cdot \frac{22}{48} = \frac{253}{9996}$

   b. $\frac{13}{52} \cdot \frac{12}{51} \cdot \frac{11}{50} \cdot \frac{10}{49} \cdot \frac{9}{48} = \frac{33}{66,640}$

   c. 0

30. $\frac{0.35}{0.65} = 0.54$

31. $\frac{0.16}{0.3} = 0.53$

32. $\frac{0.57}{0.7} = 0.81$

33. $\frac{0.028}{0.5} = 0.056$

34. a. $\frac{1}{2}$   b. $\frac{3}{7}$

35. $1 - (0.45)^6 = 0.99$

36. $1 - (\frac{5}{6})^4 = 0.518$

37. $1 - (0.15)^6 = 0.9999886$

38. 2,646

39. 40,320

40. 1,365

41. 1,188,137,600; 710,424,000

42. 720

43. 33,554,432

44. 56

45. $\frac{1}{4}$

46. $\frac{3}{14}$

47. $\frac{12}{55}$

48.

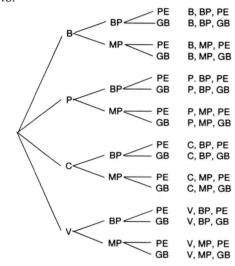

Note:  Answers may vary due to rounding, TI-83's or computer programs.

EXERCISE SET 5-1

1.
A random variable is a variable whose values are determined by chance.  Examples will vary.

3.
The number of commercials a radio station plays during each hour.
The number of times a student uses his or her calculator during a mathematics exam.
The number of leaves on a specific type of tree.

5.
A probability distribution is a distribution which consists of the values a random variable can assume along with the corresponding probabilities of these values.

7.
No; probabilities cannot be negative and the sum of the probabilities is not 1.

9.
Yes

11.
No, probability values cannot be greater than 1.

13.
Discrete

15.
Continuous

17.
Discrete

19.

| X | 0 | 1 | 2 | 3 |
|---|---|---|---|---|
| P(X) | $\frac{6}{15}$ | $\frac{5}{15}$ | $\frac{3}{15}$ | $\frac{1}{15}$ |

19. continued

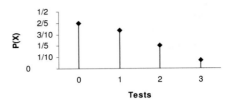

21.

| X | 2 | 3 | 5 | 7 |
|---|---|---|---|---|
| P(X) | 0.35 | 0.41 | 0.15 | 0.09 |

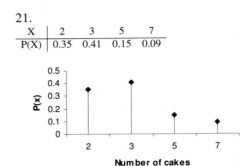

23.

| X | 1 | 2 | 3 | 4 | 5 | 6 |
|---|---|---|---|---|---|---|
| P(X) | $\frac{1}{2}$ | $\frac{1}{6}$ | $\frac{1}{12}$ | $\frac{1}{12}$ | $\frac{1}{12}$ | $\frac{1}{12}$ |

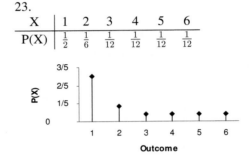

25.

| X | 2 | 3 | 4 | 5 |
|---|---|---|---|---|
| P(X) | 0.01 | 0.34 | 0.62 | 0.03 |

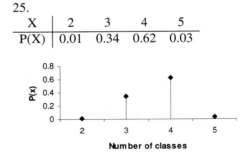

27.

| X | $1 | $5 | $10 | $20 |
|---|---|---|---|---|
| P(X) | $\frac{3}{11}$ | $\frac{2}{11}$ | $\frac{5}{11}$ | $\frac{1}{11}$ |

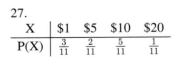

27. continued

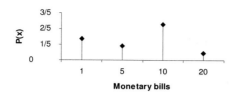

29.

| X | 1 | 2 | 3 | 4 |
|---|---|---|---|---|
| P(X) | $\frac{1}{4}$ | $\frac{1}{4}$ | $\frac{3}{8}$ | $\frac{1}{8}$ |

31.

| X | 1 | 2 | 3 |
|---|---|---|---|
| P(X) | $\frac{1}{6}$ | $\frac{1}{3}$ | $\frac{1}{2}$ |

Yes.

33.

| X | 3 | 4 | 7 |
|---|---|---|---|
| P(X) | $\frac{3}{6}$ | $\frac{4}{6}$ | $\frac{7}{6}$ |

No, the sum of the probabilities is greater than one and $P(7) = \frac{7}{6}$ which is also greater than 1.

35.

| X | 1 | 2 | 4 |
|---|---|---|---|
| P(X) | $\frac{1}{7}$ | $\frac{2}{7}$ | $\frac{4}{7}$ |

Yes.

EXERCISE SET 5-2

1.

| X | 0 | 1 | 2 | 3 |
|---|---|---|---|---|
| P(X) | 0.92 | 0.03 | 0.03 | 0.02 |

$\mu = \sum X \cdot P(X) = 0(0.92) + 1(0.03) +$

$2(0.03) + 3(0.02) = 0.15$ or 0.2

$\sigma^2 = \sum X^2 \cdot P(X) - \mu^2 = [0^2(0.92) + 1^2(0.03) + 2^2(0.03) + 3^2(0.02)] - 0.15^2 = 0.3075$ or 0.3

$\sigma = \sqrt{0.3075} = 0.55$ or 0.6

The company would need 0.2(10) = 2 extra transistors on hand each day.

1. continued

| X | P(X) | X · P(X) | X² · P(X) |
|---|---|---|---|
| 0 | 0.92 | 0 | 0 |
| 1 | 0.03 | 0.03 | 0.03 |
| 2 | 0.03 | 0.06 | 0.12 |
| 3 | 0.02 | 0.06 | 0.18 |
| | | $\mu = 0.15$ | 0.33 |

3.

$\mu = \sum X \cdot P(X) = 0(0.18) + 1(0.44) + 2(0.27) + 3(0.08) + 4(0.03) = 1.34$ or 1.3

$\sigma^2 = \sum X^2 \cdot P(X) - \mu^2 = [0^2(0.18) + 1^2(0.44) + 2^2(0.27) + 3^2(0.08) + 4^2(0.03)] - 1.34^2 = 0.92$ or 0.9

$\sigma = \sqrt{0.92} = 0.96$ or 1

No, on average each person has about one credit card.

| X | P(X) | X · P(X) | X² · P(X) |
|---|---|---|---|
| 0 | 0.18 | 0 | 0 |
| 1 | 0.44 | 0.44 | 0.44 |
| 2 | 0.27 | 0.54 | 1.08 |
| 3 | 0.08 | 0.24 | 0.72 |
| 4 | 0.03 | 0.12 | 0.48 |
| | | $\mu = 1.34$ | 2.72 |

5.

$\mu = \sum X \cdot P(X) = 4(0.4) + 5(0.3) + 6(0.1) + 8(0.15) + 10(0.05) = 5.4$

$\sigma^2 = \sum X^2 \cdot P(X) - \mu^2 = [4^2(0.4) + 5^2(0.3) + 6^2(0.1) + 8^2(0.15) + 10^2(0.05)] - 5.4^2 = 2.94$

$\sigma = \sqrt{2.94} = 1.71$

| X | P(X) | X · P(X) | X² · P(X) |
|---|---|---|---|
| 4 | 0.40 | 1.6 | 6.4 |
| 5 | 0.30 | 1.5 | 7.5 |
| 6 | 0.10 | 0.6 | 3.6 |
| 8 | 0.15 | 1.2 | 9.6 |
| 10 | 0.05 | 0.5 | 5.0 |
| | | $\mu = 5.4$ | 32.1 |

P(selling 6 or more contracts) = 0.3

7.

$\mu = \sum X \cdot P(X) = 5(0.2) + 6(0.25) + 7(0.38) + 8(0.10) + 9(0.07) = 6.59$ or 6.6

7. continued
$\sigma^2 = \sum X^2 \cdot P(X) - \mu^2 = [5^2(0.2) +$
$6^2(0.25) + 7^2(0.38) + 8^2(0.10) +$
$9^2(0.07) - 6.59^2 = 1.2619$ or $1.3$

$\sigma = \sqrt{1.2619} = 1.123$ or $1.1$

| X | P(X) | X · P(X) | X² · P(X) |
|---|---|---|---|
| 5 | 0.20 | 1.00 | 5.00 |
| 6 | 0.25 | 1.50 | 9.00 |
| 7 | 0.38 | 2.66 | 18.62 |
| 8 | 0.10 | 0.80 | 6.40 |
| 9 | 0.07 | 0.63 | 5.67 |
| | | $\mu = 6.59$ | 44.69 |

9.
$\mu = \sum X \cdot P(X) = 6(0.15) + 8(0.3) +$
$10(0.35) + 12(0.1) + 14(0.1) = 9.4$

$\sigma^2 = \sum X^2 \cdot P(X) - \mu^2 = [6^2(0.15) +$
$8^2(0.3) + 10^2(0.35) + 12^2(0.1) + 14^2(0.1)]$
$- 9.4^2 = 5.24$
$\sigma = \sqrt{5.24} = 2.289$
$P(< 8 \text{ or } > 12) = 0.15 + 0.10 = 0.25$

| X | P(X) | X · P(X) | X² · P(X) |
|---|---|---|---|
| 6 | 0.15 | 0.9 | 5.4 |
| 8 | 0.30 | 2.4 | 19.2 |
| 10 | 0.35 | 3.5 | 35.0 |
| 12 | 0.10 | 1.2 | 14.4 |
| 14 | 0.10 | 1.4 | 19.6 |
| | | $\mu = 9.4$ | 93.6 |

11.
$E(X) = \sum X \cdot P(X) = \$300(0.998) -$
$\$19,700(0.002) = \$260$

13.
$E(X) = \sum X \cdot P(X) = \$5.00(\frac{1}{6}) = \$0.83$
He should pay about $0.83.

15.
$E(X) = \sum X \cdot P(X) = \$1000(\frac{1}{1000}) +$
$\$500(\frac{1}{1000}) + \$100(\frac{5}{1000}) - \$3.00$
$= -\$1.00$

Alternate Solution:
$E(X) = 997(\frac{1}{1000}) + 497(\frac{1}{1000}) + 97(\frac{5}{1000})$
$- 3(\frac{993}{1000}) = -\$1.00$

17.
$E(X) = \sum X \cdot P(X) = \$500(\frac{1}{1000}) - \$1.00$
$= -\$0.50$

17. continued
If 123 is boxed:
$E(X) = \$499(\frac{1}{1000}) - 1(\frac{999}{1000}) = -\$0.50$

There are 6 possibilities when a number with all different digits is boxed, $(3 \cdot 2 \cdot 1 = 6)$. Hence,
$\$80.00 \cdot \frac{6}{1000} - \$1.00 = \$0.48 - \$1.00$
$= -\$0.52$

Alternate Solution:
$E(X) = 79(\frac{6}{1000}) - 1(\frac{994}{1000}) = -\$0.52$

19.
Total amount of money required: $4000
Number of tickets sold: 1000
Cost per ticket: $\frac{4000}{1000} = \$4$ per ticket

21.
The expected value for a single die is 3.5, and since 3 die are rolled, the expected value is $3(3.5) = 10.5$

23.
Answers will vary.

25.
Answers will vary.

EXERCISE SET 5-3

1.
a. Yes
b. Yes
c. Yes
d. No, there are more than two outcomes.
e. No, there are more than two outcomes.
f. Yes
g. Yes
h. Yes
i. No, there are more than two outcomes.
j. Yes

2.
a. 0.420
b. 0.346
c. 0.590
d. 0.251
e. 0.000
f. 0.250
g. 0.418
h. 0.176
i. 0.246

3.

a. $P(X) = \frac{n!}{(n-X)!\,X!} \cdot p^X \cdot q^{n-X}$

$P(X) = \frac{6!}{3!\cdot 3!} \cdot (0.03)^3(0.97)^3 = 0.0005$

b. $P(X) = \frac{4!}{2!\cdot 2!} \cdot (0.18)^2 \cdot (0.82)^2 = 0.131$

c. $P(X) = \frac{5!}{2!\cdot 3!} = (0.63)^3 \cdot (0.37)^2 = 0.342$

d. $P(X) = \frac{9!}{9!\cdot 0!} \cdot (0.42)^0 \cdot (0.58)^9 = 0.007$

e. $P(X) = \frac{10!}{5!\cdot 5!} \cdot (0.37)^5 \cdot (0.63)^5 = 0.173$

5.

n = 20, p = 0.5, X ≥ 15
$P(X) = 0.015 + 0.005 + 0.001 = 0.021$
No, it's only about a 2% chance.

7.

a. n = 8, p = 0.77, X = 8

$P(X) = \frac{8!}{0!\,8!}(0.77)^8(0.23)^0 = 0.124$

b. n = 8, P = 0.77, X = 5, 6, 7, or 8

$P(X) = \frac{8!}{3!5!}(0.77)^5(0.23)^3 +$

$\frac{8!}{2!6!}(0.77)^6(0.23)^2 + \frac{8!}{1!7!}(0.77)^7(0.23)^1 +$

$\frac{8!}{0!8!}(0.77)^8(0.23)^0 = 0.184 + 0.309 + 0.295$

$+ 0.124 = 0.912$

c. n = 8, p = 0.77, X = 3

$P(X) = \frac{8!}{5!3!}(0.77)^3(0.23)^5 = 0.017$

9.

n = 7, p = 0.75, X = 0, 1, 2, 3

$P(X) = \frac{7!}{7!\,0!}(0.75)^0(0.25)^7 +$

$\frac{7!}{6!\,1!}(0.75)^1(0.25)^6 + \frac{7!}{5!\,2!}(0.75)^2(0.25)^5 +$

$\frac{7!}{4!\,3!}(0.75)^3(0.25)^4 = 0.071$

11.

n = 5, p = 0.40
a. X = 2, P(X) = 0.346
b. X = 0, 1, 2, or 3 people
$P(X) = 0.078 + 0.259 + 0.346 + 0.230$
$= 0.913$

11. continued

c. X = 2, 3, 4, or 5 people
$P(X) = 0.346 + 0.230 + 0.077 + 0.01$
$= 0.663$
d. X = 0, 1, or 2 people
$P(X) = 0.683$

13.

a. n = 10, p = 0.53, X = 5

$P(X) = \frac{10!}{5!5!}(0.53)^5(0.47)^5 = 0.2417$

b. n = 10, p = 0.47, X ≥ 5

$P(X) = \frac{10!}{5!5!}(0.47)^5(0.53)^5 +$

$\frac{10!}{6!4!}(0.47)^6(0.53)^4 + \frac{10!}{7!3!}(0.47)^7(0.53)^3 +$

$\frac{10!}{8!2!}(0.47)^8(0.53)^2 + \frac{10!}{9!1!}(0.47)^9(0.53)^1 +$

$\frac{10!}{10!0!}(0.47)^{10}(0.53)^0$

$P(X \geq 5) = 0.548$

c. n = 10, p = 0.53, X < 5  — less than 5

$P(X) = \frac{10!}{5!5!}(0.53)^5(0.47)^5 +$

$\frac{10!}{4!6!}(0.53)^4(0.47)^6 + \frac{10!}{3!7!}(0.53)^3(0.47)^7 +$

$\frac{10!}{2!8!}(0.53)^2(0.47)^8 + \frac{10!}{1!9!}(0.53)^1(0.47)^9$

$\frac{10!}{0!10!}(0.53^0)(0.47)^{10} = 0.306$

14.

a. $\mu = 100(0.75) = 75$
$\sigma^2 = 100(0.75)(0.25) = 18.75$ or $18.8$
$\sigma = \sqrt{18.75} = 4.33$ or $4.3$
b. $\mu = 300(0.3) = 90$
$\sigma^2 = 300(0.3)(0.7) = 63$
$\sigma = \sqrt{63} = 7.94$ or $7.9$
c. $\mu = 20(0.5) = 10$
$\sigma^2 = 20(0.5)(0.5) = 5$
$\sigma = \sqrt{5} = 2.236$ or $2.2$
d. $\mu = 10(0.8) = 8$
$\sigma^2 = 10(0.8)(0.2) = 1.6$
$\sigma = \sqrt{1.6} = 1.265$ or $1.3$
e. $\mu = 1000(0.1) = 100$
$\sigma^2 = 1000(0.1)(0.9) = 90$
$\sigma = \sqrt{90} = 9.49$ or $9.5$

**14. continued**

f. $\mu = 500(0.25) = 125$

$\sigma^2 = 500(0.25)(0.75) = 93.75$ or $93.8$

$\sigma = \sqrt{93.75} = 9.68$ or $9.7$

g. $\mu = 50(\frac{2}{5}) = 20$

$\sigma^2 = 50(\frac{2}{5})(\frac{3}{5}) = 12$

$\sigma = \sqrt{12} = 3.464$ or $3.5$

h. $\mu = 36(\frac{1}{6}) = 6$

$\sigma^2 = 36(\frac{1}{6})(\frac{5}{6}) = 5$

$\sigma = \sqrt{5} = 2.236$ or $2.2$

**15.**

$n = 800, p = 0.01$

$\mu = 800(0.01) = 8$

$\sigma^2 = 800(0.01)(0.99) = 7.9$

$\sigma = \sqrt{7.92} = 2.8$

**17.**

$n = 300, p = 0.03$

$\mu = 300(0.03) = 9$

$\sigma^2 = 300(0.03)(0.97) = 8.73$

$\sigma = \sqrt{8.73} = 2.95$

**19.**

$n = 1000, p = 0.21$

$\mu = 1000(0.21) = 210$

$\sigma^2 = 1000(0.21)(0.79) = 165.9$

$\sigma = \sqrt{165.9} = 12.9$

**21.**

$n = 18, p = 0.25, X = 5$

$P(X) = \frac{18!}{13!\,5!}(0.25)^5(0.75)^{13} = 0.199$

**23.**

$n = 10, p = \frac{1}{3}, X = 0, 1, 2, 3$

$P(X) = \frac{10!}{10!\,0!}(\frac{1}{3})^0(\frac{2}{3})^{10} + \frac{10!}{9!\,1!}(\frac{1}{3})^1(\frac{2}{3})^9$

$+ \frac{10!}{8!\,2!}(\frac{1}{3})^2(\frac{2}{3})^8 + \frac{10!}{7!\,3!}(\frac{1}{3})^3(\frac{2}{3})^7 = 0.559$

**25.**

$n = 20, p = 0.58, X = 12$

$P(X) = \frac{20!}{8!\,12!}(0.58)^{12}(0.42)^8 = 0.177$

**27.**

$n = 7, p = 0.14, X = 2$ or $3$

$P(X) = \frac{7!}{5!\,2!}(0.14)^2(0.86)^5 +$

$\frac{7!}{4!\,3!}(0.14)^3(0.86)^4 = 0.246$

**29.**

| X | 0 | 1 | 2 | 3 |
|---|---|---|---|---|
| P(X) | 0.125 | 0.375 | 0.375 | 0.125 |

**EXERCISE SET 5-4**

**1.**

a. $P(M) = \frac{6!}{3!\,2!\,1!}(0.5)^3(0.3)^2(0.2)^1 = 0.135$

b. $P(M) = \frac{5!}{1!\,2!\,2!}(0.3)^1(0.6)^2(0.1)^2 = 0.0324$

c. $P(M) = \frac{4!}{1!\,1!\,2!}(0.8)^1(0.1)^1(0.1)^2 = 0.0096$

d. $P(M) = \frac{3!}{1!\,1!\,1!}(0.5)^1(0.3)^1(0.2)^1 = 0.18$

e. $P(M) = \frac{5!}{1!\,3!\,1!}(0.7)^1(0.2)^3(0.1)^1 = 0.0112$

**3.**

$P(M) =$

$\frac{12!}{2!\,2!\,2!\,2!\,2!\,2!}(0.13)^2(0.13)^2(0.14)^2(0.16)^2(0.2)^2$

$(0.24)^2 = 0.00247$

**5.**

$P(M) = \frac{4!}{2!\,1!\,1!}(\frac{1}{6})^2(\frac{1}{6})^1(\frac{1}{6})^1 = \frac{1}{108}$

**7.**

a. $P(5; 4) = 0.1563$

b. $P(2; 4) = 0.1465$

c. $P(6; 3) = 0.0504$

d. $P(10; 7) = 0.071$

e. $P(9; 8) = 0.1241$

**9.**

$p = \frac{1}{20,000} = 0.00005$

$\lambda = n \cdot p = 80,000(0.00005) = 4$

a. $P(0; 4) = 0.0183$

b. $P(1; 4) = 0.0733$

c. $P(2; 4) = 0.1465$

d. $P(3 \text{ or more}; 4) = 1 - [P(0; 4) + P(1; 4)$

$+ P(2; 4)]$

$= 1 - (0.0183 + 0.0733 + 0.1465)$

$= 0.7619$

**11.**

$p = \frac{5}{1000} = \frac{1}{200}$

$\lambda = n \cdot p = (250) \cdot (\frac{1}{200}) = 1.25$

$P(\text{at least 2 orders}) =$

$1 - [P(0 \text{ orders} + P(1 \text{ order})]$

$= 1 - [\frac{e^{-1.25}(1.25)^0}{0!} + \frac{e^{-1.25}(1.25)^1}{1!}]$

$= 1 - (0.2865 + 0.3581) = 0.3554$

**13.**

$\lambda = 200(0.015) = 3$

$P(0; 3) = 0.0498$

**15.**

$P(5; 4) = 0.1563$

**17.**

P(one from each

class) $= \frac{_5C_1 \cdot _4C_1 \cdot _5C_1 \cdot _7C_1}{_{21}C_4} = \frac{700}{5985} = 0.117$

**19.**

P(2 jazz and 1

classical) $= \frac{_{10}C_2 \cdot _4C_1 \cdot _2C_0}{_{16}C_3} = \frac{180}{560} = 0.321$

**21.**

P(at least 1 defective) $= 1 - P(0 \text{ defectives})$

$a = 6, b = 18, n = 3, X = 0$

$P(0) = \frac{_6C_0 \cdot _{18}C_3}{_{24}C_3} = \frac{102}{253} = 0.403$

P(at least 1 defective) $= 1 - 0.403 = 0.597$

## REVIEW EXERCISES - CHAPTER 5

**1.**

Yes.

**3.**

No, the sum of the probabilities is greater than 1.

**5.**

| X | 0 | 1 | 2 | 3 | 4 | 5 |
|------|------|------|------|------|------|------|
| P(X) | 0.27 | 0.28 | 0.2 | 0.15 | 0.08 | 0.02 |

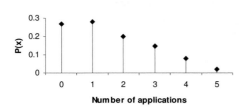

a. P(2 or 3

applications) $= 0.2 + 0.15 = 0.35$

b.

$\mu = 0(0.27) + 1(0.28) + 2(0.2) + 3(0.15) + 4(0.08) + 5(0.02) = 1.55$

$\sigma^2 = [0^2(0.27) + 1^2(0.28) + 2^2(0.2) + 3^2(0.15) + 4^2(0.08) + 5^2(0.02)] - 1.55^2 = 1.8075$

$\sigma = \sqrt{1.8075} = 1.3444$

**5b. continued**

| X | P(X) | X · P(X) | X² · P(X) |
|---|------|----------|-----------|
| 0 | 0.27 | 0 | 0 |
| 1 | 0.28 | 0.28 | 0.28 |
| 2 | 0.2 | 0.40 | 0.80 |
| 3 | 0.15 | 0.45 | 1.35 |
| 4 | 0.08 | 0.32 | 1.28 |
| 5 | 0.02 | 0.10 | 0.50 |
| | | $\mu = 1.55$ | 4.21 |

**7.**

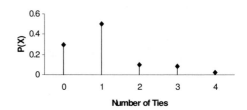

**9.**

$\mu = \sum X \cdot P(X) = 13(0.12) + 14(0.15) + 15(0.29) + 16(0.25) + 17(0.19) = 15.24$ or 15.2

$\sigma^2 = \sum X^2 \cdot P(X) - \mu^2 = [13^2(0.12) + 14^2(0.15) + 15^2(0.29) + 16^2(0.25) + 17^2(0.19)] - 15.24^2 = 1.5824$ or 1.6

$\sigma = \sqrt{1.5824} = 1.26$ or 1.3

| X | P(X) | X · P(X) | X² · P(X) |
|----|------|----------|-----------|
| 13 | 0.12 | 1.56 | 20.28 |
| 14 | 0.15 | 2.1 | 29.4 |
| 15 | 0.29 | 4.35 | 65.25 |
| 16 | 0.25 | 4 | 64 |
| 17 | 0.19 | 3.23 | 54.91 |
| | | $\mu = 15.24$ | 233.84 |

**11.**

$\mu = \sum X \cdot P(X) = 22(0.08) + 23(0.19) + 24(0.36) + 25(0.25) + 26(0.07) + 27(0.05) = 24.19$ or 24.2

$\sigma^2 = \sum X^2 \cdot P(X) - \mu^2 = [22^2(0.08) + 23^2(0.19) + 24^2(0.36) + 25^2(0.25) + 26^2(0.07) + 27^2(0.05)] - 24.19^2 = 1.4539$ or 1.5

$\sigma = \sqrt{1.4539} = 1.206$ or 1.2

11. continued

| X | P(X) | X · P(X) | $X^2 \cdot P(X)$ |
|----|------|------|------|
| 22 | 0.08 | 1.76 | 38.72 |
| 23 | 0.19 | 4.37 | 100.51 |
| 24 | 0.36 | 8.64 | 207.36 |
| 25 | 0.25 | 6.25 | 156.25 |
| 26 | 0.07 | 1.82 | 47.32 |
| 27 | 0.05 | 1.35 | 36.45 |
|  | $\mu =$ | 24.19 | 586.61 |

13.
Let $x$ = cost to play the game
$P(ace) = \frac{1}{13}$     $P(face\ card) = \frac{3}{13}$
$P(2 - 10) = \frac{9}{13}$
For a fair game, $E(X) = 0$.
$0 = -20(\frac{1}{13}) + 10(\frac{3}{13}) + 2(\frac{9}{13}) - x$
$x = \$2.15$

15.
n = 12, p = 0.3
a.  $P(X = 8) = 0.008$
b.  $P(X < 5) = 0.724$
c.  $P(X \geq 10) = 0.0002$
d.  $P(4 < x \leq 9) = 0.2761$

17.
$\mu = n \cdot p = 180(0.75) = 135$
$\sigma^2 = n \cdot p \cdot q = 180(0.75)(0.25) = 33.75$ or
33.8
$\sigma = \sqrt{33.75} = 5.809$ or 5.8

19.
n = 8, p = 0.25
$P(X \leq 3) = \frac{8!}{8!\,0!}(0.25)^0(0.75)^8 +$
$\frac{8!}{7!\,1!}(0.25)^1(0.75)^7 + \frac{8!}{6!\,2!}(0.25)^2(0.75)^6 +$
$\frac{8!}{5!\,3!}(0.25)^3(0.75)^5 = 0.8862$ or 0.886

21.
n = 20, p = 0.75, X = 16
P(16 have eaten pizza for breakfast) =

$\frac{20!}{4!\,16!}(0.75)^{16}(0.25)^4 = 0.1897$ or 0.190

23.
$P(M) = \frac{10!}{5!\,3!\,1!\,1!}(0.46)^5(0.41)^3(0.09)^1(0.03)^1$
$= 0.0193$

25.
$P(M) = \frac{10!}{5!\,3!\,2!}(0.50)^5(0.40)^3(0.10)^2 = 0.050$

27.
a.  P(6 or more; 6) = 1 − P(5 or less; 6)
= 1 − (0.0025 + 0.0149 + 0.0446 +
0.0892 + 0.1339 + 0.1606) = 0.5543
b.  P(4 or more; 6) = 1 − P(3 or less; 6)
= 1 − (0.0025 + 0.0149 + 0.0446 +
0.0892) = 0.8488
c.  P(5 or less; 6) = P(0; 6) + ... + P(6; 6)
= 0.4457

29.
a = 13, b = 39, n = 5, X = 2
$P(2) = \frac{_{13}C_2 \cdot _{39}C_3}{_{52}C_5} = \frac{9,139}{33,320} = 0.27$

31.
P(1 vegetable & 2 fruit) =

$\frac{_{10}C_0 \cdot _8C_1 \cdot _8C_2}{_{26}C_3} = \frac{224}{2600} = 0.0862$

CHAPTER 5 QUIZ

1.  True
2.  False, it is a discrete random variable.
3.  False, the outcomes must be independent.
4.  True
5.  chance
6.  $\mu = n \cdot p$
7.  one
8.  c.
9.  c.
10. d.
11. No, the sum of the probabilities is
greater than one.
12. Yes
13. Yes
14. Yes
15.

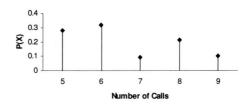

16.

| X | 0 | 1 | 2 | 3 | 4 |
|------|------|------|------|------|------|
| P(X) | 0.02 | 0.30 | 0.48 | 0.13 | 0.07 |

16. continued

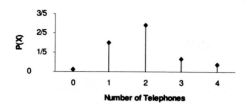

17.

$\mu = 0(0.10) + 1(0.23) + 2(0.31) + 3(0.27)$
$+ 4(0.09) = 2.02$ or 2
$\sigma^2 = [0^2(0.10) + 1^2(0.23) + 2^2(0.31) +$
$3^2(0.27) + 4^2(0.09)] - 2.02^2 = 1.3$
$\sigma = \sqrt{1.3} = 1.1$

18.

$\mu = 30(0.05) + 31(0.21) + 32(0.38) +$
$33(0.25) + 34(0.11) = 32.16$ or 32.2
$\sigma^2 = [30^2(0.05) + 31^2(0.21) + 32^2(0.38) +$
$33^2(0.25) + 34^2(0.11)] - 32.16^2 = 1.07$ or
1.1
$\sigma = \sqrt{1.07} = 1.0$

19.

$\mu = 4(\frac{1}{6}) + 5(\frac{1}{6}) + 2(\frac{1}{6}) + 10(\frac{1}{6}) + 3(\frac{1}{6})$
$+ 7(\frac{1}{6}) = 5.17$ or 5.2

20.

$\mu = \$2(\frac{1}{2}) + \$10(\frac{5}{26}) + \$25(\frac{3}{26}) +$
$\$100(\frac{1}{26}) = \$9.65$

21.

n = 20, p = 0.40, X = 5
P(5 ) = 0.124

22.

n = 20, p = 0.60
a. P(15) = 0.075
b. P(10, 11, ..., 20) = 0.872
c. P(0, 1, 2, 3, 4, 5) = 0.125

23.

n = 300, p = 0.80
$\mu = 300(0.80) = 240$
$\sigma^2 = 300(0.80)(0.20) = 48$
$\sigma = \sqrt{48} = 6.9$

24.

n = 75, p = 0.12
$\mu = 75(0.12) = 9$
$\sigma^2 = 75(0.12)(0.88) = 7.9$
$\sigma = \sqrt{7.9} = 2.8$

25.

$P(M) = \frac{30!}{15!\,8!\,5!\,2!}(0.5)^{15}(0.3)^8(0.15)^5(0.05)^2$

$= 0.0080$

26.

$P(M) = \frac{16!}{9!\,4!\,3!}(0.88)^9(0.08)^4(0.04)^3$

$= 0.0003$

27.

$P(M) = \frac{12!}{5!\,4!\,3!}(0.45)^5(0.35)^4(0.2)^3$

$= 0.061$

28.

$\lambda = 100(0.08) = 8, \ X = 6$
P(6; 8) = 0.122

29.

$\lambda = 8$
a. $P(X \geq 8; 8) = 0.1396 + \ldots + 0.0001$
$= 0.5471$
b. $P(X \geq 3; 8) = 1 - P(0, 1, \text{ or } 2 \text{ calls})$
$= 1 - (0.0003 + 0.0027 + 0.0107)$
$= 1 - 0.0137 = 0.9863$
c. $P(X \leq 7; 8) = 0.0003 + \ldots + 0.1396$
$= 0.4529$

30.

a = 12, b = 36, n = 6, X = 3

$P(A) = \frac{_{12}C_3 \cdot _{36}C_3}{_{48}C_6} = \frac{\frac{12!}{9!\,3!} \cdot \frac{36!}{33!\,3!}}{\frac{48!}{42!\,6!}} = 0.128$

31.

a. $\frac{_6C_3 \cdot _8C_1}{_{14}C_4} = \frac{\frac{6!}{3!\,3!} \cdot \frac{8!}{7!\,1!}}{\frac{14!}{10!\,4!}} = 0.16$

b. $\frac{_6C_2 \cdot _8C_2}{_{14}C_4} = \frac{\frac{6!}{4!\,2!} \cdot \frac{8!}{6!\,2!}}{\frac{14!}{10!\,4!}} = 0.42$

c. $\frac{_6C_0 \cdot _8C_4}{_{14}C_4} = \frac{\frac{6!}{6!\,0!} \cdot \frac{8!}{4!\,4!}}{\frac{14!}{10!\,4!}} = 0.07$

Note to instructors: Graphs are not to scale and are intended to convey a general idea.

Answers are generated using Table E. Answers generated using the TI-83 will vary slightly.

EXERCISE SET 6-1

1.
The characteristics of the normal distribution are:
1. It is bell-shaped.
2. It is symmetric about the mean.
3. The mean, median, and mode are equal.
4. It is continuous.
5. It never touches the X-axis.
6. The area under the curve is equal to one.
7. It is unimodal.

3.
One or 100%.

5.
68%, 95%, 99.7%

7.
The area is found by looking up $z = 0.75$ in Table E and subtracting 0.5.
Area $= 0.7734 - 0.5 = 0.2734$

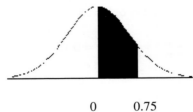

0     0.75

9.
The area is found by looking up $z = -2.07$ in Table E and subtracting from 0.5.
Area $= 0.5 - 0.0192 = 0.4808$

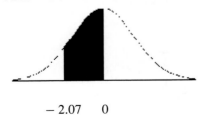

$-2.07$   0

11.
The area is found by looking up $z = 0.23$ in Table E and subtracting it from 1.

11. continued
Area $= 1 - 0.5910 = 0.4090$

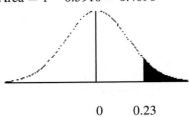

0     0.23

13.
The area is found by looking up $z = -1.43$ in Table E.
Area $= 0.0764$

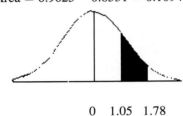

$-1.43$    0

15.
The area is found by looking up the values 1.05 and 1.78 in Table E and subtracting the areas.
Area $= 0.9625 - 0.8531 = 0.1094$

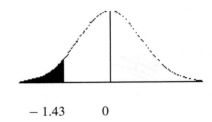

0  1.05  1.78

17.
The area is found by looking up the values $-1.56$ and $-1.83$ in Table E and subtracting the areas.
Area $= 0.0594 - 0.0336 = 0.0258$

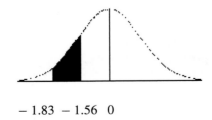

$-1.83$ $-1.56$  0

**19.**
The area is found by looking up the values
$-1.53$ and $2.08$ in Table E and subtracting
the areas.
Area $= 0.9812 - 0.0630 = 0.9182$

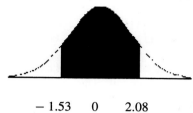

$-1.53 \quad 0 \quad 2.08$

**21.**
The area is found by looking up $z = 2.11$ in
Table E.
Area $= 0.9826$

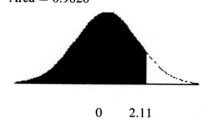

$0 \quad 2.11$

**23.**
The area is found by looking up $-0.25$ in
Table E and subtracting it from 1.
$1 - 0.4013 = 0.5987$

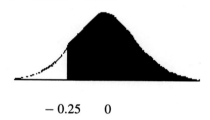

$-0.25 \quad 0$

**25.**
For $z = -0.44$, the area is $0.3300$. For
$z = 1.92$, the area is $1 - 0.9726 = 0.0274$
Area $= 0.3300 + 0.0274 = 0.3574$

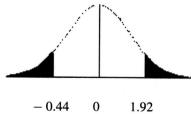

$-0.44 \quad 0 \quad 1.92$

**27.**
Area $= 0.7486 - 0.5 = 0.2486$

**27. continued**

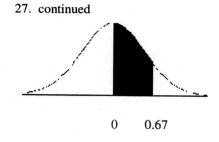

$0 \quad 0.67$

**29.**
Area $= 0.5 - 0.0582 = 0.4418$

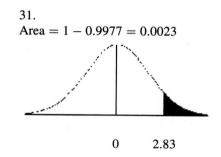

$-1.57 \quad 0$

**31.**
Area $= 1 - 0.9977 = 0.0023$

$0 \quad 2.83$

**33.**
Area $= 0.1131$

$-1.21 \quad 0$

**35.**
Area $= 0.9591 - 0.0069 = 0.9522$
(TI answer $= 0.9521$)

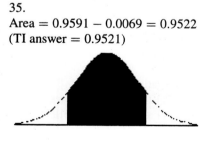

$-2.46 \quad 0 \quad 1.74$

**37.**
Area = 0.9985 − 0.9279 = 0.0706
(TI answer = 0.0707)

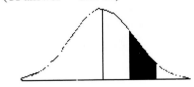

0  1.46  2.97

**39.**
Area = 0.9222

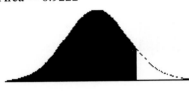

0    1.42

**41.**
Since the $z$ score is on the left side of 0, use the negative $z$ table. Areas in the negative $z$ table are in the tail, so we will use 0.5 − 0.4175 = 0.0825 as the area. The closest $z$ score corresponding to an area of 0.0825 is $z = -1.39$.
(TI answer = −1.3885)

**43.**
$z = -2.08$, found by using the negative $z$ table. (TI answer = −2.0792)

**45.**
Use the negative $z$ table and 1 − 0.8962 = 0.1038 for the area. The $z$ score is $z = -1.26$.
(TI answer = −1.2602)

**47.**
a. Using the negative $z$ table, area = 1 − 0.9887 = 0.0113. Hence $z = -2.28$. (TI answer = −2.2801)

b. Using the negative $z$ table, area = 1 − 0.8212 = 0.1788. Hence $z = -0.92$. (TI answer = −0.91995)

c. Using the negative $z$ table, area = 1 − 0.6064 = 0.3936. Hence $z = -0.27$. (TI answer = −0.26995)

**49.**
a. For total area = 0.05, there will be area = 0.025 in each tail. The $z$ scores are ± 1.96. (TI answer = ± 1.95996)

b. For total area = 0.10, there will be area = 0.05 in each tail. The $z$ scores are $z = \pm 1.645$. (TI answer = ± 1.64485)

c. For total area = 0.01, there will be area = 0.005 in each tail. The $z$ scores are $z = \pm 2.58$. (TI answer = ± 2.57583)

**51.**
$P(-1 < z < 1) = 0.8413 - 0.1587$
$= 0.6826$

$P(-2 < z < 2) = 0.9772 - 0.0228$
$= 0.9544$

$P(-3 < z < 3) = 0.9987 - 0.0013$
$= 0.9974$

They are very close.

**53.**
For $z = -1.2$, area = 0.1151
Area (left side) = 0.5 − 0.1151 = 0.3849
0.8671 − 0.3849 = 0.4822
Area (right side) = 0.4822 + 0.5 = 0.9822
For area = 0.9822, $z = 2.10$
Thus, $P(-1.2 < z < 2.10) = 0.8671$

**55.**
For $z = -0.5$, area = 0.3085
0.3085 − 0.2345 = 0.074
For area = 0.074, $z = -1.45$
Thus, $P(-1.45 < z < -0.5) = 0.2345$

For $z = -0.5$, area = 0.3085
0.5 − 0.3085 = 0.1915
0.2345 − 0.1915 = 0.043
0.5 + 0.043 = 0.543
For area = 0.543, $z = 0.11$
Thus, $P(-0.5 < z < 0.11) = 0.2345$

**57.**
$$y = \frac{e^{\frac{-(X-0)^2}{2(1)^2}}}{1\sqrt{2\pi}} = \frac{e^{\frac{-X^2}{2}}}{\sqrt{2\pi}}$$

## EXERCISE SET 6-2

1.
$$z = \frac{\$3.50 - \$5.81}{\$0.81} = -2.85$$

$P(z < -2.85) = 0.0022$ or $0.22\%$

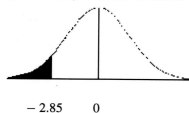

$-2.85 \qquad 0$

3.
a. $z = \frac{750,000 - 706,242}{52,145} = 0.84$

$P(z > 0.84) = 1 - 0.7995 = 0.2005$ or
$20.05\%$
(TI answer $= 0.2007$)

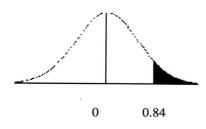

$0 \qquad 0.84$

b. $z = \frac{600,000 - 706,242}{52,145} = -2.04$

$z = \frac{700,000 - 706,242}{52,145} = -0.12$

$P(-2.04 < z < -0.12) = 0.4522 - 0.0207$
$P = 0.4315$ or $43.15\%$
(TI answer $= 0.4316$)

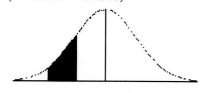

$-2.04 \quad -0.12$

5.
a. $z = \frac{200 - 225}{10} = -2.5$

$z = \frac{220 - 225}{10} = -0.5$

$P(-2.5 < z < -0.5) =$
$0.3085 - 0.0062 = 0.3023$ or $30.23\%$

5a. continued

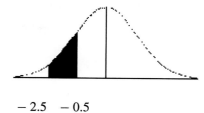

$-2.5 \quad -0.5$

b. $z = -2.5$

$P(z < -2.5) = 0.0062$ or $0.62\%$

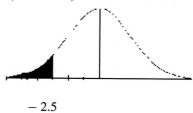

$-2.5$

7.
a. $z = \frac{\$90,000 - \$85,900}{\$11,000} = 0.37$

$P(z > 0.37) = 1 - 0.6443 = 0.3557$
or $35.57\%$
(TI answer $= 0.3547$)

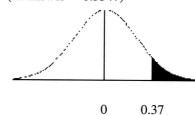

$0 \qquad 0.37$

b. $z = \frac{\$75,000 - \$85,900}{\$11,000} = -0.99$

$P(z > -0.99) = 1 - 0.1611$
$= 0.8389$ or $83.89\%$
(TI answer $= 0.8391$)

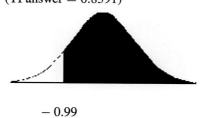

$-0.99$

9.
For $x \geq 15,000$ miles:

$z = \frac{15,000 - 12,494}{1290} = 1.94$

9. continued
$$P(z > 1.94) = 1 - 0.9738 = 0.0262$$
(TI answer = 0.02603)

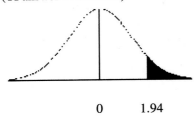

For $x < 8000$ miles:

$$z = \frac{8000 - 12,494}{1290} = -3.48$$

$$P(z < -3.48) = 0.0003$$
(TI answer = 0.00025)

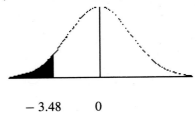

For $x < 6000$ miles:

$$z = \frac{6000 - 12,494}{1290} = -5.03$$

$$P(z < -5.03) = 0.0001$$

Maybe it would be good to know why it had only been driven less than 6000 miles.

11.
a. $z = \frac{1000 - 3262}{1100} = -2.06$

$$P(z \geq -2.06) = 1 - 0.0197 = 0.9803 \text{ or } 98.03\%$$
(TI answer = 0.9801)

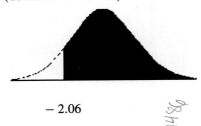

b. $z = \frac{4000 - 3262}{1100} = 0.67$

$$P(z > 0.67) = 1 - 0.7486 = 0.2514 \text{ or } 25.14\%$$
(TI answer = 0.2511)

11b. continued

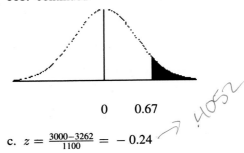

c. $z = \frac{3000 - 3262}{1100} = -0.24$ ⟶ .4052

$$P(-0.24 < z < 0.67) =$$
$$0.7486 - 0.4052 = 0.3434 \text{ or } 34.34\%$$
(TI answer = 0.3430)

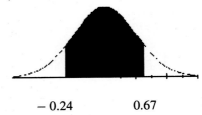

13.
a. $z = \frac{350 - 380}{16} = -1.88$

$$P(z \geq -1.88) = 1 - 0.0301 = 0.9699 \text{ or } 96.99\%$$
(TI answer = 0.9696)

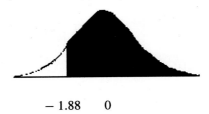

b. $z = \frac{395 - 380}{16} = 0.94$

$$P(z \leq 0.94) = 0.8264$$
(TI answer = 0.8257)

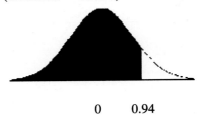

c. Use the range rule of thumb:
If $\frac{\text{Range}}{4} \approx 16$, then the range is $16 \cdot 4 = 64$.

15.

a. $z = \frac{15-23.5}{3.6} = -2.36$

$z = \frac{22-23.5}{3.6} = -0.42$

$P(-2.36 < z < -0.42) =$
$0.3372 - 0.0091 = 0.3281$

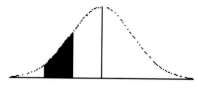

−2.36  −0.42

b. $z = \frac{18-23.5}{3.6} = -1.53$

$z = \frac{25-23.5}{3.6} = 0.42$

$P(z < -1.53 \text{ or } z > 0.42) =$
$0.0630 + (1 - 0.6628) =$
$0.063 + 0.3372 = 0.4002$

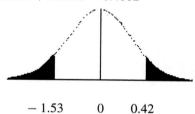

−1.53   0   0.42

c. For 15 minutes, $z = -2.36$.
$P(z < -2.36) = 0.0091$
Since the probability is small, it is not likely
that a person would be seated in less than 15
minutes.

17.
The middle 66% means that 33% of the area
will be on either side of the mean. Thus,
area = 0.33 and $z = \pm 0.95$.
$x = -0.95(1025) + 6492 = \$5518.25$
$x = 0.95(1025) + 6492 = \$7465.75$

The prices are between $5518.25 and
$7465.75.
(TI answers: $5513.98 < \mu < \$7470.02$)

$5518.25        $7465.75

17. continued
Yes, a boat priced at $5550 would be sold in
this store.

19.
The middle 80% means that 40% of the area
will be on either side of the mean. The
corresponding $z$ scores will be $\pm 1.28$.
$x = -1.28(92) + 1810 = 1692.24$ sq. ft.
$x = 1.28(92) + 1810 = 1927.76$ sq. ft.
(TI answers: 1927.90 maximum, 1692.10
minimum)

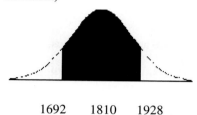

1692      1810     1928

21.
$z = \frac{1200-949}{100} = 2.51$

$P(z > 2.51) = 1 - 0.9940 = 0.006$ or $0.6\%$

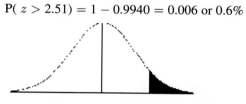

0      2.51

For the least expensive 10%, the area is 0.4
on the left side of the curve. Thus,
$z = -1.28$.
$x = -1.28(100) + 949 = \$821$

23.
The middle 60% means that 30% of the area
will be on either side of the mean. The
corresponding $z$ scores will be $\pm 0.84$.
$x = -0.84(1150) + 8256 = \$7290$
$x = 0.84(1150) + 8256 = \$9222$
(TI answer: $7288.14 < \mu < \$9223.86$)

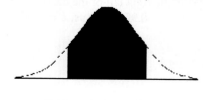

$7290   $8256   $9222

**25.**
For the longest 10%, the area is 0.90. Thus,
$z = 1.28$
Since $\sigma^2 = 2.1$, $\sigma = \sqrt{2.1} = 1.449$
$x = 1.28(1.449) + 4.8$
$x = 6.65$ or 6.7 days
(TI answer $= 6.657$)
For the shortest 30%, the area is 0.30. Thus,
$z = -0.52$.
$x = -0.52(1.449) + 4.8$
$x = 4.047$ days or 4.05 days
(TI answer $= 4.040$)

**27.**
The bottom 18% area is 0.18. Thus,
$z = -0.92$.
$x = -0.92(6256) + 24,596 = \$18,840.48$
(TI answer $= \$18,869.48$)

**29.**
The 10% to be exchanged would be at the left, or bottom, of the curve; therefore, area $= 0.10$. The corresponding $z$ score will be $-1.28$.
$x = -1.28(5) + 25 = 18.6$ months.

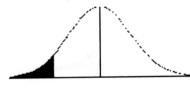

      18.6      25

**31.**
a. $\mu = 120$    $\sigma = 20$
b. $\mu = 15$    $\sigma = 2.5$
c. $\mu = 30$    $\sigma = 5$

**33.**
There are several mathematical tests that can be used including drawing a histogram and calculating Pearson's index of skewness.

**35.**
2.87% area in the right tail of the curve means that area $= 0.9713$. Thus, $z = 1.90$.

Since $z = \frac{X - \mu}{\sigma}$:
$1.90 = \frac{112 - 110}{\sigma}$
$1.90\sigma = 2$
$\sigma = 1.05$

**37.**
1.25% of the area in each tail means that area $= 0.0125$ in the left tail and area $= 0.9875$ in the right tail. Thus, $z = \pm 2.24$.
Then $\mu = \frac{42 + 48}{2} = 45$ and $x = \mu + z\sigma$.
$48 = 45 + 2.24\sigma$
$\sigma = 1.34$

**39.**
Histogram:

The histogram shows a positive skew.

$PI = \frac{3(970.2 - 853.5)}{376.5} = 0.93$

$IQR = Q_3 - Q_1 = 910 - 815 = 95$
$1.5(IQR) = 1.5(95) = 142.5$
$Q_1 - 142.5 = 672.5$
$Q_3 + 142.5 = 1052.5$
There are several outliers.

Conclusion: The distribution is not normal.

**41.**
Histogram:

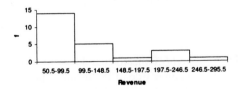

The histogram shows a positive skew.

$PI = \frac{3(115.3 - 92.5)}{66.32} = 1.03$

$IQR = Q_3 - Q_1 = 154.5 - 67 = 87.5$
$1.5(IQR) = 1.5(87.5) = 131.25$
$Q_1 - 131.25 = -64.25$
$Q_3 + 131.25 = 285.75$
There is one outlier.

Conclusion: The distribution is not normal.

EXERCISE SET 6-3

**1.**
The distribution is called the sampling distribution of sample means.

**3.**
The mean of the sample means is equal to the population mean.

**5.**
The distribution will be approximately normal when sample size is large.

**7.**
$$z = \frac{\overline{X} - \mu}{\sigma/\sqrt{n}}$$

**9.**
a. $z = \frac{\overline{X} - \mu}{\frac{\sigma}{\sqrt{n}}} = \frac{\$25,000 - \$26,489}{\frac{\$3204}{\sqrt{36}}} = -2.79$

$P(z < -2.79) = 0.0026$ or $0.26\%$
(TI answer = 0.0026)

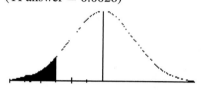

$\$25,000 \quad \$26,489$

b. $z = \frac{\overline{X} - \mu}{\frac{\sigma}{\sqrt{n}}} = \frac{\$26,000 - \$26,489}{\frac{\$3204}{\sqrt{36}}} = -0.92$

$P(z > -0.92) = 1 - 0.1788 = 0.8212$
(TI answer = 0.8201)

$26,000 \quad 26,489$

c. $z = \frac{\overline{X} - \mu}{\frac{\sigma}{\sqrt{n}}} = \frac{\$24,000 - \$26,489}{\frac{\$3204}{\sqrt{36}}} = -4.66$

$z = \frac{\overline{X} - \mu}{\frac{\sigma}{\sqrt{n}}} = \frac{\$26,000 - \$26,489}{\frac{\$3204}{\sqrt{36}}} = -0.92$

$P(-4.66 < z < -0.92) =$
$0.1788 - 0.0001 = 0.1787$
(TI answer = 0.1799)

9c. continued

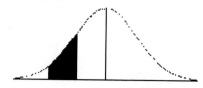

$\$24,000 \quad \$26,000$

**11.**
$z = \frac{\overline{X} - \mu}{\frac{\sigma}{\sqrt{n}}} = \frac{144.5 - 142}{\frac{12.3}{\sqrt{36}}} = 1.22$

$P(z > 1.22) = 1 - 0.8888 = 0.1112$ or $11.12\%$
(TI answer = 0.1113)

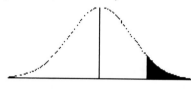

$142 \quad 144.5$

No, since the average weight of the group is within 2 standard deviations (standard errors) of the mean.

**13.**
For $x < 20$ mpg:

$z = \frac{\overline{X} - \mu}{\frac{\sigma}{\sqrt{n}}} = \frac{20 - 21}{\frac{2.9}{\sqrt{25}}} = -1.72$

$P(z < -1.72) = 0427$ or $4.27\%$
(TI answer = 0.0423)

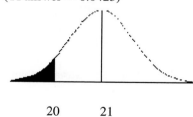

$20 \quad 21$

For $20 < x < 25$:

$z = \frac{\overline{X} - \mu}{\frac{\sigma}{\sqrt{n}}} = \frac{25 - 21}{\frac{2.9}{\sqrt{25}}} = 6.90$

$z = \frac{\overline{X} - \mu}{\frac{\sigma}{\sqrt{n}}} = \frac{20 - 21}{\frac{2.9}{\sqrt{25}}} = -1.72$

13. continued
$P(20 < z < 25) = 0.9999 - 0.0427$
$= 0.9572$ or $95.72\%$
(TI answer $= 0.9577$)

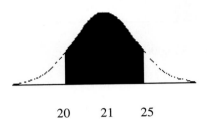

15.
a. $z = \frac{\overline{X}-\mu}{\sigma} = \frac{670-660}{35} = 0.29$

$P(z > 0.29) = 1 - 0.6141 = 0.3859$
(TI answer $= 0.3875$)

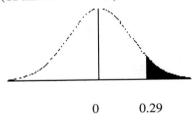

b. $z = \frac{\overline{X}-\mu}{\frac{\sigma}{\sqrt{n}}} = \frac{670-660}{\frac{35}{\sqrt{10}}} = 0.90$

$P(z > 0.90) = 1 - 0.8159 = 0.1841$
(TI answer $= 0.1831$)

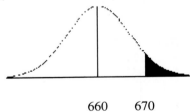

c. Individual values are more variable than means.

17.
$z = \frac{\overline{X}-\mu}{\frac{\sigma}{\sqrt{n}}} = \frac{120-123}{\frac{21}{\sqrt{15}}} = -0.55$

$z = \frac{\overline{X}-\mu}{\frac{\sigma}{\sqrt{n}}} = \frac{126-123}{\frac{21}{\sqrt{15}}} = 0.55$

$P(-0.55 < z < 0.55) = 0.7088 - 0.2912$
$= 0.4176$ or $41.76\%$
(TI answer $= 0.4199$)

17. continued

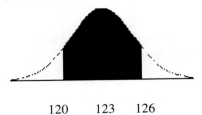

120    123    126

19.
$z = \frac{\overline{X}-\mu}{\frac{\sigma}{\sqrt{n}}} = \frac{1980-2000}{\frac{187.5}{\sqrt{50}}} = -0.75$

$z = \frac{\overline{X}-\mu}{\frac{\sigma}{\sqrt{n}}} = \frac{1990-2000}{\frac{187.5}{\sqrt{50}}} = -0.38$

$P(-0.75 < z < -0.38) =$
$0.3520 - 0.2266 = 0.1254$
(TI answer $= 0.12769$)

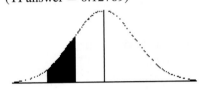

1980  1990  2000

21.
a. $z = \frac{\overline{X}-\mu}{\sigma} = \frac{43-46.2}{8} = -0.4$

$P(z < -0.4) = 0.3446$ or $34.46\%$

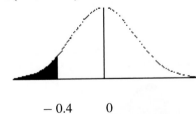

$-0.4$    0

b. $z = \frac{43-46.2}{\frac{8}{\sqrt{50}}} = -2.83$

$P(z < -2.83) = 0.0023$ or $0.23\%$

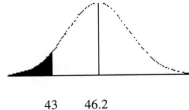

43    46.2

c. Yes, since it is within one standard deviation of the mean.

21. continued
d. Very unlikely, since the probability would be less than 1%.

23.
a. $z = \frac{220 - 215}{15} = 0.33$

$P(z > 0.33) = 1 - 0.6293 = 0.3707$ or 37.07%
(TI answer = 0.3694)

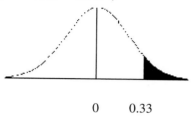

$$0 \qquad 0.33$$

b. $z = \frac{220 - 215}{\frac{15}{\sqrt{25}}} = 1.67$

$P(z > 1.67) = 1 - 0.9525 = 0.0475$ or 4.75%
(TI answer = 0.04779)

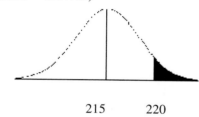

$$215 \qquad 220$$

25.
$z = \frac{800 - 777}{\frac{77}{\sqrt{50}}} = 2.11$

$P(z > 2.11) = 1 - 0.9826 = 0.0174$

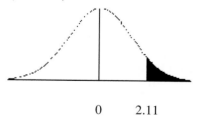

$$0 \qquad 2.11$$

27.
Since $50 > 0.05(800)$ or 40, the correction factor is necessary.

It is $\sqrt{\frac{800 - 50}{800 - 1}} = 0.969$

$z = \frac{\overline{X} - \mu}{\frac{\sigma}{\sqrt{n}} \cdot \sqrt{\frac{N-n}{n-1}}} = \frac{83,500 - 82,000}{\frac{5000}{\sqrt{50}}(0.969)} = 2.19$

27. continued
$P(z > 2.19) = 1 - 0.9857 = 0.0143$ or 1.43%

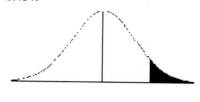

$$82,000 \qquad 83,500$$

29.
$\sigma_x = \frac{\sigma}{\sqrt{n}} = \frac{15}{\sqrt{100}} = 1.5$

$2(1.5) = \frac{15}{\sqrt{n}}$

$3 \cdot \sqrt{n} = 15$

$\sqrt{n} = 5$

$n = 25$, the sample size necessary to double the standard error.

EXERCISE SET 6-4

1.
When p is approximately 0.5, and as n increases, the shape of the binomial distribution becomes similar to the normal distribution. The normal approximation should be used only when $n \cdot p$ and $n \cdot q$ are both greater than or equal to 5. The correction for continuity is necessary because the normal distribution is continuous and the binomial is discrete.

2.
For each problem use the following formulas:
$\mu = np \qquad \sigma = \sqrt{npq} \qquad z = \frac{\overline{X} - \mu}{\sigma}$
Be sure to correct each X for continuity.
a. $\mu = 0.5(30) = 15$
$\sigma = \sqrt{(0.5)(0.5)(30)} = 2.74$

$z = \frac{17.5 - 15}{2.74} = 0.91 \qquad$ area = 0.8186

$z = \frac{18.5 - 15}{2.74} = 1.28 \qquad$ area = 0.8997

$P(17.5 < X < 18.5) = 0.8997 - 0.8186$
$= 0.0811 = 8.11\%$

2. continued

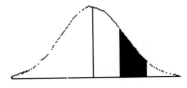

15    17.5   18.5

b. $\mu = 0.8(50) = 40$
$\sigma = \sqrt{(50)(0.8)(0.2)} = 2.83$

$z = \frac{43.5-40}{2.83} = 1.24$      area $= 0.8925$

$z = \frac{44.5-40}{2.83} = 1.59$      area $= 0.9441$

$P(43.5 < X < 44.5) = 0.9441 - 0.8925$
$= 0.0516$ or $5.16\%$

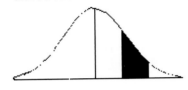

40    43.5  44.5

c. $\mu = 0.1(100) = 10$
$\sigma = \sqrt{(0.1)(0.9)(100)} = 3$

$z = \frac{11.5-10}{3} = 0.50$      area $= 0.6915$

$z = \frac{12.5-10}{3} = 0.83$      area $= 0.7967$

$P(11.5 < X < 12.5) = 0.7967 - 0.6915$
$= 0.1052$ or $10.52\%$

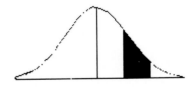

10    11.5  12.5

d. $\mu = 10(0.5) = 5$
$\sigma = \sqrt{(0.5)(0.5)(10)} = 1.58$

$z = \frac{6.5-5}{1.58} = 0.95$      area $= 0.8289$

$P(X \geq 6.5) = 1 - 0.8289 = 0.1711$ or
$17.11\%$

2d. continued

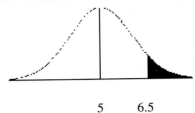

5        6.5

e. $\mu = 20(0.7) = 14$
$\sigma = \sqrt{(20)(0.7)(0.3)} = 2.05$

$z = \frac{12.5-14}{2.05} = -0.73$      area $= 0.2327$

$P(X \leq 12.5) = 0.2327$ or $23.27\%$

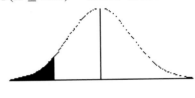

12.5       14

f. $\mu = 50(0.6) = 30$
$\sigma = \sqrt{(50)(0.6)(0.4)} = 3.46$

$z = \frac{40.5-30}{3.46} = 3.03$      area $= 0.9988$

$P(X \leq 40.5) = 0.9988$ or $99.88\%$

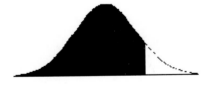

30       40.5

3.
a. np $= 20(0.50) = 10 \geq 5$     Yes
   nq $= 20(0.50) = 10 \geq 5$
b. np $= 10(0.60) = 6 \geq 5$     No
   nq $= 10(0.40) = 4 < 5$
c. np $= 40(0.90) = 36 \geq 5$     No
   nq $= 40(0.10) = 4 < 5$
d. np $= 50(0.20) = 10 \geq 5$     Yes
   nq $= 50(0.80) = 40 \geq 5$
e. np $= 30(0.80) = 24 \geq 5$     Yes
   nq $= 30(0.20) = 6 \geq 5$
f. np $= 20(0.85) = 17 \geq 5$     No
   nq $= 20(0.15) = 3 \ < 5$

**5.**

$p = \frac{2}{5} = 0.4 \qquad \mu = 400(0.4) = 160$

$\sigma = \sqrt{(400)(0.4)(0.6)} = 9.8$

$z = \frac{169.5 - 160}{9.8} = 0.97 \qquad\qquad \text{area} = 0.8340$

$P(X > 169.5) = 1 - 0.8340 = 0.1660$ or 16.6%

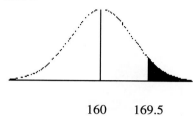

160     169.5

**7.**

$\mu = 300(0.531) = 159.3$

$\sigma = \sqrt{(300)(0.531)(0.469)} = 8.64$

$z = \frac{175.5 - 159.3}{8.64} = 1.88 \text{ area} = 0.9699$

$P(X > 175.5) = 1 - 0.9699 = 0.0301$
(TI answer = 0.0304)

153.9     175.5

**9.**

$\mu = 200(0.261) = 52.2$

$\sigma = \sqrt{(200)(0.261)(0.739)} = 6.21$

$z = \frac{49.5 - 52.2}{6.21} = -0.43 \quad \text{area} = 0.3336$

$P(X \geq 49.5) = 1 - 0.3336 = 0.6664$
(TI answer = 0.6681)

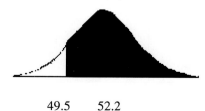

49.5     52.2

**11.**

$\mu = 300(0.803) = 240.9$

$\sigma = \sqrt{(300)(0.803)(0.197)} = 6.89$

$X > \frac{3}{4}(300)$ or $X > 225$

**11. continued**

$z = \frac{225.5 - 240.9}{6.89} = -2.24 \quad \text{area} = 0.0125$

$P(X > 225.5) = 1 - 0.0125 = 0.9875$
(TI answer = 0.9873)

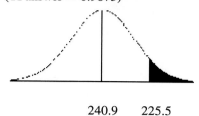

240.9     225.5

**13.**

$\mu = 350(0.35) = 122.5$

$\sigma = \sqrt{(350)(0.35)(0.65)} = 8.92$

$z = \frac{99.5 - 122.5}{8.92} = -2.58 \qquad \text{area} = 0.0049$

$P(X > 99.5) = 1 - 0.0049 = 0.9951$ or
99.51% (TI answer = 0.9950)

Yes; it is likely that 100 or more people would favor the parking lot.

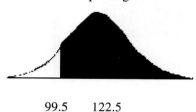

99.5     122.5

**15.**

a. $n(0.1) = 5 \qquad\qquad n \geq 50$
b. $n(0.3) = 5 \qquad\qquad n \geq 17$
c. $n(0.5) = 5 \qquad\qquad n \geq 10$
d. $n(0.2) = 5 \qquad\qquad n \geq 25$
e. $n(0.1) = 5 \qquad\qquad n \geq 50$

**REVIEW EXERCISES - CHAPTER 6**

**1.**
a. $0.9744 - 0.5 = 0.4744$

0     1.95

1. continued
b. $0.6443 - 0.5 - 0.1443$

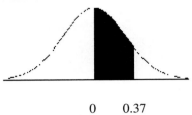

0    0.37

c. $0.9656 - 0.9066 = 0.0590$

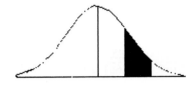

0   1.32  1.82

d. $0.9798 - 0.1469 = 0.8329$
(TI answer $= 0.83296$)

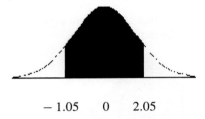

$-1.05$    0    2.05

e. $0.7019 - 0.4880 = 0.2139$

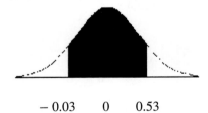

$-0.03$    0    0.53

f. $0.8643 - 0.0359 = 0.8284$

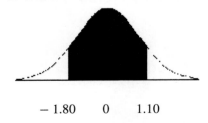

$-1.80$    0    1.10

g. $1 - 0.9767 = 0.0233$

1g. continued

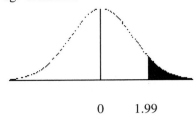

0      1.99

h. $1 - 0.0869 = 0.9131$

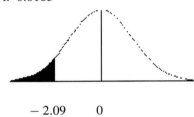

$-1.36$    0

i. $0.0183$

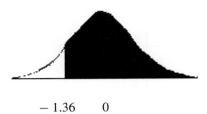

$-2.09$    0

j. $0.9535$

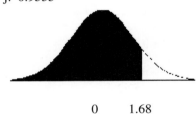

0      1.68

3.
a. $z = \frac{\$6000 - \$5274}{\$600} = 1.21$

$P(z > 1.21) = 1 - 0.8869 = 0.1131$

0      1.21

b. For the middle 50%, 25% of the area is on each side of 0. Thus, $z = \pm 0.67$

3b. continued
$x = 0.67(600) + 5274 = \$5676$
$x = -0.67(600) + 5274 = \$4872$
(TI answers: \$4869.31 to \$5678.69)

5.
For exceeding the speed limit (X > 65):

$z = \frac{65-63}{8} = 0.25$

$P(z > 0.25) = 1 - 0.5987 = 0.4013$ or
40.13%

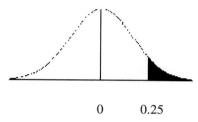

For exceeding 72 mph (X > 72):

$z = \frac{72-63}{8} = 1.13$

$P(z > 1.13) = 1 - 0.8708 = 0.1292$ or
12.92% might be arrested.
(TI answer = 0.1303 or 13.03%)

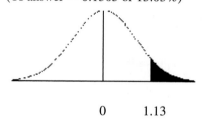

7.
a. $z = \frac{18 - 19.32}{2.44} = -0.54$

$P(z > -0.54) = 1 - 0.2946 = 0.7054$
(TI answer = 0.7057)

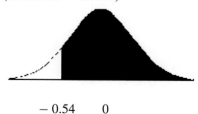

b. $z = \frac{18 - 19.32}{\frac{2.44}{\sqrt{5}}} = -1.21$

$P(z > -1.21) = 1 - 0.1131 = 0.8869$
(TI answer = 0.8868)

7b. continued

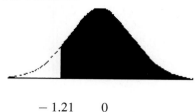

−1.21    0

9.
a. $z = \frac{X-\mu}{\frac{\sigma}{\sqrt{n}}} = \frac{27 - 25.7}{\frac{3.75}{\sqrt{40}}} = 2.19$

$P(\overline{X} > 27) = 1 - 0.9857 = 0.0143$
(TI answer = 0.0142)

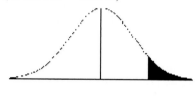

25.7    27

b. $z = \frac{X-\mu}{\frac{\sigma}{\sqrt{n}}} = \frac{\$60 - \$61.50}{\frac{\$5.89}{\sqrt{50}}} = -1.80$

$P(\overline{X} > 60) = 1 - 0.0359 = 0.9641$

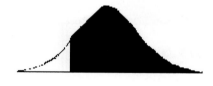

\$60    \$61.50

11.
$z = \frac{3.4-3.7}{\frac{0.6}{\sqrt{32}}} = -2.83$

$P(\overline{X} < 3.4) = 1 - 0.9977 = 0.0023$ or
0.23%

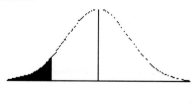

3.4    3.7

Yes, since the probability is less than 1%.

13.
For fewer than 10 holding multiple jobs:
$\mu = 150(0.053) = 7.95$

13. continued
$$\sigma = \sqrt{(150)(0.053)(0.947)} = 2.744$$

$$z = \frac{9.5 - 7.95}{2.74} = 0.56$$

$P(X < 9.5) = 0.7123$
(TI answer $= 0.7139$)

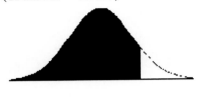

7.95        9.5

For more than 50 <u>not</u> holding multiple jobs:
$\mu = 150(0.947) = 142.05$
$\sigma = \sqrt{150(0.947)(0.053)} = 2.744$

$$z = \frac{50.5 - 142.05}{2.744} = -33.37$$

$P(X > 50.5) = 1 - 0.0001 = 0.9999$
(TI answer $= 0.9999$)

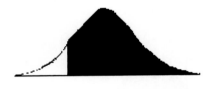

50.5    142.05

15.
$\mu = 200(0.2) = 40$
$\sigma = \sqrt{(200)(0.2)(0.8)} = 5.66$

$$z = \frac{49.5 - 40}{5.66} = 1.68$$

$P(X \geq 49.5) = 1 - 0.9535 = 0.0465$ or
$4.65\%$

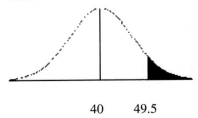

40        49.5

17.
Histogram:

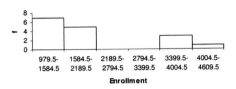

The histogram shows a positive skew.

$$PI = \frac{3(2136.1 - 1755)}{1171.7} = 0.98$$

$IQR = Q_3 - Q_1$
$IQR = 2827 - 1320 = 1507$
$1.5(IQR) = 1.5(1507) = 2260.5$
$Q_1 - 2260.5 = -940.5$
$Q_3 + 2260.5 = 5087.5$

There are no outliers.

Conclusion: The distribution is not normal.

CHAPTER 6 QUIZ

1. False, the total area is equal to one.
2. True
3. True
4. True
5. False, the area is positive.
6. False, it applies to means taken from the same population.
7. a.
8. a.
9. b.
10. b.
11. c.
12. 0.5
13. sampling error
14. the population mean
15. the standard error of the mean
16. 5
17. 5%

18. the areas are:
a. 0.4332    f. 0.8284
b. 0.3944    g. 0.0401
c. 0.0344    h. 0.8997
d. 0.1029    i. 0.017
e. 0.2912    j. 0.9131

19. the probabilities are:
a. 0.4846   f. 0.0384
b. 0.4693   g. 0.0089
c. 0.9334   h. 0.9582
d. 0.0188   i. 0.9788
e. 0.7461   j. 0.8461

20. the probabilities are:
a. 0.7734
b. 0.0516
c. 0.3837
d. Any rainfall above 65 inches could be considered an extremely wet year since this value is two standard deviations above the mean.

21. the probabilities are:
a. 0.0668   c. 0.4649
b. 0.0228   d. 0.0934

22. the probabilities are:
a. 0.4525   c. 0.3707
b. 0.3707   d. 0.019

23. the probabilities are:
a. 0.0013   c. 0.0081
b. 0.5      d. 0.5511

24. the probabilities are:
a. 0.0037   c. 0.5
b. 0.0228   d. 0.3232

25. 8.804 cm
26. The lowest acceptable score is 121.24.
27. 0.015
28. 0.9738
29. 0.0495; no
30. 0.0630
31. 0.8577
32. 0.0499
33. The distribution is not normal.
34. The distribution is approximately normal.

Note: Answers may vary due to rounding.

EXERCISE SET 7-1

**1.**
A point estimate of a parameter specifies a specific value such as $\mu = 87$, whereas an interval estimate specifies a range of values for the parameter such as $84 < \mu < 90$. The advantage of an interval estimate is that a specific confidence level (say 95%) can be selected, and one can be 95% confident that the parameter being estimated lies in the interval.

**3.**
The maximum error of estimate is the likely range of values to the right or left of the statistic in which may contain the parameter.

**5.**
A good estimator should be unbiased, consistent, and relatively efficient.

**7.**
To determine sample size, the maximum error of estimate and the degree of confidence must be specified and the population standard deviation must be known.

**9.**
a. 2.58     d. 1.65
b. 2.33     e. 1.88
c. 1.96

**11.**
a. $\overline{X} = 82$ is the point estimate for $\mu$.

b. $\overline{X} - z_{\frac{\alpha}{2}}\left(\frac{s}{\sqrt{n}}\right) < \mu < \overline{X} + z_{\frac{\alpha}{2}}\left(\frac{s}{\sqrt{n}}\right)$
$82 - (1.96)\left(\frac{15}{\sqrt{35}}\right) < \mu < 82 + (1.96)\left(\frac{15}{\sqrt{35}}\right)$
$82 - 4.97 < \mu < 82 + 4.97$
$77 < \mu < 87$

c. $82 - (2.58)\left(\frac{15}{\sqrt{35}}\right) < \mu < 82 + (2.58)\left(\frac{15}{\sqrt{35}}\right)$
$82 - 6.54 < \mu < 82 + 6.54$
$75 < \mu < 89$
(TI answer: $75.5 < \mu < 88.5$)

d. The 99% confidence interval is larger because the confidence level is larger.

**13.**
$\overline{X} - z_{\frac{\alpha}{2}}\left(\frac{\sigma}{\sqrt{n}}\right) < \mu < \overline{X} + z_{\frac{\alpha}{2}}\left(\frac{\sigma}{\sqrt{n}}\right)$
$1.8 - 1.65\left(\frac{0.33}{\sqrt{50}}\right) < \mu < 1.8 + 1.65\left(\frac{0.33}{\sqrt{50}}\right)$
$1.8 - 0.08 < \mu < 1.8 + 0.08$
$1.72 < \mu < 1.88$

The estimate is lower.

**15.**
$\overline{X} - z_{\frac{\alpha}{2}}\left(\frac{s}{\sqrt{n}}\right) < \mu < \overline{X} + z_{\frac{\alpha}{2}}\left(\frac{s}{\sqrt{n}}\right)$
$\$150,000 - 1.96\left(\frac{15,000}{\sqrt{35}}\right) < \mu <$
$\$150,000 + 1.96\left(\frac{15,000}{\sqrt{35}}\right)$
$\$150,000 - 4969.51 < \mu <$
$\$150,000 + 4969.51$

$\$145,030 < \mu < \$154,970$
(TI answer: $\$145,031 < \mu < \$154,969$)

**17.**
$\overline{X} = 5000 \quad s = 900$
$5000 - 1.96\left(\frac{900}{\sqrt{415}}\right) < \mu < 5000 +$
$1.96\left(\frac{900}{\sqrt{415}}\right)$
$4913 < \mu < 5087$

4000 hours does not seem reasonable since it is outside this interval.

**19.**
$\overline{X} - z_{\frac{\alpha}{2}}\left(\frac{s}{\sqrt{n}}\right) < \mu < \overline{X} + z_{\frac{\alpha}{2}}\left(\frac{s}{\sqrt{n}}\right)$
$61.2 - 1.96\left(\frac{7.9}{\sqrt{84}}\right) < \mu < 61.2 + 1.96\left(\frac{7.9}{\sqrt{84}}\right)$
$61.2 - 1.69 < \mu < 61.2 + 1.69$
$59.5 < \mu < 62.9$

**21.**
$n = \left[\frac{z_{\frac{\alpha}{2}} \cdot \sigma}{E}\right]^2 = \left[\frac{(2.58)(6.2)}{1.5}\right]^2$
$n = (10.664)^2 = 113.7$ or 114

**23.**
$n = \left[\frac{z_{\frac{\alpha}{2}} \cdot \sigma}{E}\right]^2 = \left[\frac{(2.33)(10.1)}{2}\right]^2$
$n = (11.77)^2 = 138.5$ or 139

**25.**
$n = \left[\frac{z_{\frac{\alpha}{2}} \cdot \sigma}{E}\right]^2 = \left[\frac{(2.58)(30)}{5}\right]^2$
$n = (15.48)^2 = 239.6$ or 240

**EXERCISE SET 7-2**

**1.**

The characteristics of the t-distribution are: It is bell-shaped, symmetrical about the mean, and never touches the x-axis. The mean, median, and mode are equal to 0 and are located at the center of the distribution. The variance is greater than 1. The t-distribution is a family of curves based on degrees of freedom. As sample size increases the t-distribution approaches the normal distribution.

**3.**

The t-distribution should be used when $\sigma$ is unknown and n < 30.

**4.**

a. 2.898 where d. f. $= 17$
b. 2.074 where d. f. $= 22$
c. 2.624 where d. f. $= 14$
d. 1.833 where d. f. $= 9$
e. 2.093 where d. f. $= 19$

**5.**

$\overline{X} - t_{\frac{\alpha}{2}}(\frac{s}{\sqrt{n}}) < \mu < \overline{X} + t_{\frac{\alpha}{2}}(\frac{s}{\sqrt{n}})$

$16 - (2.861)(\frac{2}{\sqrt{20}}) < \mu < 16 + (2.861)(\frac{2}{\sqrt{20}})$

$16 - 1.28 < \mu < 16 + 1.28$

$15 < \mu < 17$

**7.**

$\overline{X} = 33.4 \quad s = 28.7$

$\overline{X} - t_{\frac{\alpha}{2}}(\frac{s}{\sqrt{n}}) < \mu < \overline{X} + t_{\frac{\alpha}{2}}(\frac{s}{\sqrt{n}})$

$33.4 - 1.746(\frac{28.7}{\sqrt{17}}) < \mu < 33.4 + 1.746(\frac{28.7}{\sqrt{17}})$

$33.4 - 12.2 < \mu < 33.4 + 12.2$

$21.2 < \mu < 45.6$

The point estimate is 33.4 and is close to the actual population mean of 32, which is within the 90% confidence interval. The mean may not be the best estimate since the data value 132 is large and possibly an outlier.

**9.**

$\overline{X} - t_{\frac{\alpha}{2}}(\frac{s}{\sqrt{n}}) < \mu < \overline{X} + t_{\frac{\alpha}{2}}(\frac{s}{\sqrt{n}})$

$276 - 2.015(\frac{12}{\sqrt{6}}) < \mu <$
$\quad 276 + 2.015(\frac{12}{\sqrt{6}})$

$276 - 9.9 < \mu < 276 + 9.9$

$266.1 < \mu < 285.9$ or $266 < \mu < 286$

The coach's claim is highly unlikely.

**11.**

$\overline{X} - t_{\frac{\alpha}{2}}(\frac{s}{\sqrt{n}}) < \mu < \overline{X} + t_{\frac{\alpha}{2}}(\frac{s}{\sqrt{n}})$

$14.3 - 2.052(\frac{2}{\sqrt{28}}) < \mu < 14.3 + 2.052(\frac{2}{\sqrt{28}})$

$14.3 - 0.8 \; < \mu < 14.3 + 0.8$

$13.5 < \mu < 15.1$

The employees should allow about 30 minutes.

**13.**

$\overline{X} - t_{\frac{\alpha}{2}}(\frac{s}{\sqrt{n}}) < \mu < \overline{X} + t_{\frac{\alpha}{2}}(\frac{s}{\sqrt{n}})$

$19.2 - 2.201(\frac{2.1}{\sqrt{12}}) < \mu <$
$\quad 19.2 + 2.201(\frac{2.1}{\sqrt{12}})$

$19.2 - 1.33 < \mu < 19.2 + 1.33$

$17.87 < \mu < 20.53$

The estimate is higher than the national average.

**15.**

$\overline{X} - t_{\frac{\alpha}{2}}(\frac{s}{\sqrt{n}}) < \mu < \overline{X} + t_{\frac{\alpha}{2}}(\frac{s}{\sqrt{n}})$

$115 - 2.571(\frac{6}{\sqrt{6}}) < \mu < 115 + 2.571(\frac{6}{\sqrt{6}})$

$115 - 6.298 < \mu < 115 + 6.298$

$109 < \mu < 121$

**17.**

$\overline{X} = 51.47 \quad s = 45.98$

$\overline{X} - t_{\frac{\alpha}{2}}(\frac{s}{\sqrt{n}}) < \mu < \overline{X} + t_{\frac{\alpha}{2}}(\frac{s}{\sqrt{n}})$

$51.47 - 1.746(\frac{45.98}{\sqrt{17}}) < \mu <$
$\quad 51.47 + 1.746(\frac{45.98}{\sqrt{17}})$

$51.47 - 19.47 < \mu < 51.47 + 19.47$

$32.0 < \mu < 70.9$

**19.**

Answers will vary.

**21.**

$\overline{X} = 2.175 \quad s = 0.585$

For $\mu > \overline{X} - t_{\frac{\alpha}{2}}(\frac{s}{\sqrt{n}})$:

$\mu > 2.175 - 1.729(\frac{0.585}{\sqrt{20}})$

$\mu > 2.175 - 0.226$

Thus, $\mu > \$1.95$ means that one can be 95% confident that the mean revenue is greater than $1.95.

For $\mu < \overline{X} + t_{\frac{\alpha}{2}}(\frac{s}{\sqrt{n}})$:

$\mu < 2.175 + 1.729(\frac{0.585}{\sqrt{20}})$

$\mu < 2.175 + 0.226$

Thus, $\mu < \$2.40$ means that one can be 95% confident that the mean revenue is less than $2.40.

EXERCISE SET 7-3

1.
a. $\hat{p} = \frac{40}{80} = 0.5$ $\qquad$ $\hat{q} = \frac{40}{80} = 0.5$

b. $\hat{p} = \frac{90}{200} = 0.45$ $\qquad$ $\hat{q} = \frac{110}{200} = 0.55$

c. $\hat{p} = \frac{60}{130} = 0.46$ $\qquad$ $\hat{q} = \frac{70}{130} = 0.54$

d. $\hat{p} = \frac{35}{60} = 0.58$ $\qquad$ $\hat{q} = \frac{25}{60} = 0.42$

e. $\hat{p} = \frac{43}{95} = 0.45$ $\qquad$ $\hat{q} = \frac{52}{95} = 0.55$

2.
For each part, change the percent to a decimal by dividing by 100, and find $\hat{q}$ using $\hat{q} = 1 - \hat{p}$.
a. $\hat{p} = 0.15$ $\qquad$ $\hat{q} = 1 - 0.15 = 0.85$
b. $\hat{p} = 0.37$ $\qquad$ $\hat{q} = 1 - 0.37 = 0.63$
c. $\hat{p} = 0.71$ $\qquad$ $\hat{q} = 1 - 0.71 = 0.29$
d. $\hat{p} = 0.51$ $\qquad$ $\hat{q} = 1 - 0.51 = 0.49$
e. $\hat{p} = 0.79$ $\qquad$ $\hat{q} = 1 - 0.79 = 0.21$

3.
$\hat{p} = 0.39$ $\qquad$ $\hat{q} = 0.61$
$\hat{p} - (z_{\frac{\alpha}{2}})\sqrt{\frac{\hat{p}\hat{q}}{n}} < p < \hat{p} + (z_{\frac{\alpha}{2}})\sqrt{\frac{\hat{p}\hat{q}}{n}}$
$0.39 - (1.96)\sqrt{\frac{(0.39)(0.61)}{1500}} < p <$
$\qquad 0.39 + (1.96)\sqrt{\frac{(0.39)(0.61)}{1500}}$
$0.39 - 0.025 < p < 0.39 + 0.025$
$0.365 < p < 0.415$

5.
$\hat{p} = \frac{X}{n} = \frac{55}{450} = 0.12$
$\hat{q} = 1 - 0.12 = 0.88$
$\hat{p} - (z_{\frac{\alpha}{2}})\sqrt{\frac{\hat{p}\hat{q}}{n}} < p < \hat{p} + (z_{\frac{\alpha}{2}})\sqrt{\frac{\hat{p}\hat{q}}{n}}$
$0.12 - 1.96\sqrt{\frac{(0.12)(0.88)}{450}} < p < 0.12 + 1.96\sqrt{\frac{(0.12)(0.88)}{450}}$
$0.12 - 0.03 < p < 0.12 + 0.03$
0.09 or 9% $< p <$ 0.15 or 15%
(TI answer: $0.092 < p < 0.153$)
11% is contained in the confidence interval.

7.
$\hat{p} = 0.84$ $\qquad$ $\hat{q} = 0.16$
$\hat{p} - (z_{\frac{\alpha}{2}})\sqrt{\frac{\hat{p}\hat{q}}{n}} < p < \hat{p} + (z_{\frac{\alpha}{2}})\sqrt{\frac{\hat{p}\hat{q}}{n}}$
$0.84 - 1.65\sqrt{\frac{(0.84)(0.16)}{200}} < p <$
$\qquad 0.84 + 1.65\sqrt{\frac{(0.84)(0.16)}{200}}$
$0.84 - 0.043 < p < 0.84 + 0.043$
$0.797 < p < 0.883$

9.
$\hat{p} = 0.65$ $\qquad$ $\hat{q} = 0.36$
$\hat{p} - (z_{\frac{\alpha}{2}})\sqrt{\frac{\hat{p}\hat{q}}{n}} < p < \hat{p} + (z_{\frac{\alpha}{2}})\sqrt{\frac{\hat{p}\hat{q}}{n}}$
$0.65 - 1.96\sqrt{\frac{(0.65)(0.35)}{300}} < p <$
$\qquad 0.65 + 1.96\sqrt{\frac{(0.65)(0.35)}{300}}$
$0.65 - 0.054 < p < 0.65 + 0.054$
$0.596 < p < 0.704$

11.
$\hat{p} = \frac{36}{85} = 0.424$ $\quad$ $\hat{q} = \frac{49}{85} = 0.576$
$\hat{p} - (z_{\frac{\alpha}{2}})\sqrt{\frac{\hat{p}\hat{q}}{n}} < p < \hat{p} + (z_{\frac{\alpha}{2}})\sqrt{\frac{\hat{p}\hat{q}}{n}}$
$0.424 - 2.58\sqrt{\frac{(0.424)(0.576)}{85}} < p <$
$\qquad 0.424 + 2.58\sqrt{\frac{(0.424)(0.576)}{85}}$
$0.424 - 0.138 < p < 0.424 + 0.138$
$0.286 < p < 0.562$
It would not be considered larger since 0.52 is in the interval.

13.
$\hat{p} = 0.44975$ $\qquad$ $\hat{q} = 0.55025$
$\hat{p} - (z_{\frac{\alpha}{2}})\sqrt{\frac{\hat{p}\hat{q}}{n}} < p < \hat{p} + (z_{\frac{\alpha}{2}})\sqrt{\frac{\hat{p}\hat{q}}{n}}$
$0.44975 - 1.96\sqrt{\frac{(0.44975)(0.55025)}{1005}} < p <$
$\qquad 0.44975 + 1.96\sqrt{\frac{(0.44975)(0.55025)}{1005}}$
$0.44975 - 0.03076 < p <$
$\qquad 0.44975 + 0.03076$
$0.419 < p < 0.481$

15.
a. $\hat{p} = 0.25$ $\qquad$ $\hat{q} = 0.75$
$n = \hat{p}\,\hat{q}\left[\frac{z_{\frac{\alpha}{2}}}{E}\right]^2 = (0.25)(0.75)\left[\frac{2.58}{0.02}\right]^2$
$n = 3120.1875$ or 3121

b. $\hat{p} = 0.5$ $\qquad$ $\hat{q} = 0.5$
$n = \hat{p}\,\hat{q}\left[\frac{z_{\frac{\alpha}{2}}}{E}\right]^2 = (0.5)(0.5)\left[\frac{2.58}{0.02}\right]^2$
$n = 4160.25$ or 4161

17.
a. $\hat{p} = 0.25$ $\qquad$ $\hat{q} = 0.75$
$n = \hat{p}\,\hat{q}\left[\frac{z_{\frac{\alpha}{2}}}{E}\right]^2 = (0.25)(0.75)\left[\frac{1.96}{0.03}\right]^2$
$n = 800.33$ or 801

b. $\hat{p} = 0.5$ $\qquad$ $\hat{q} = 0.5$
$n = \hat{p}\,\hat{q}\left[\frac{z_{\frac{\alpha}{2}}}{E}\right]^2 = (0.5)(0.5)\left[\frac{1.96}{0.03}\right]^2$
$n = 1067.11$ or 1068

**19.**

$\hat{p} = 0.5 \qquad \hat{q} = 0.5$

$n = \hat{p}\,\hat{q}\left[\dfrac{z_{\frac{\alpha}{2}}}{E}\right]^2$

$n = (0.5)(0.5)\left[\dfrac{1.65}{0.025}\right]^2$

$n = 1089$

**21.**

$600 = (0.5)(0.5)\left[\dfrac{z}{0.04}\right]^2$

$600 = 156.25z^2$

$3.84 = z^2$

$\sqrt{3.84} = 1.96 = z$

1.96 corresponds to a 95% degree of confidence.

## EXERCISE SET 7-4

**1.**

$\chi^2$

**3.**

|    | $\chi^2_{\text{left}}$ | $\chi^2_{\text{right}}$ |
|----|----------|-----------|
| a. | 3.816  | 21.920 |
| b. | 10.117 | 30.144 |
| c. | 13.844 | 41.923 |
| d. | 0.412  | 16.750 |
| e. | 26.509 | 55.758 |

**5.**

$\dfrac{(n-1)s^2}{\chi^2_{\text{right}}} < \sigma^2 < \dfrac{(n-1)s^2}{\chi^2_{\text{left}}}$

$\dfrac{16(10.1)^2}{28.845} < \sigma^2 < \dfrac{16(10.1)^2}{6.908}$

$56.6 < \sigma^2 < 236.3$

$7.5 < \sigma < 15.4$

**7.**

Use the Range Rule of Thumb for $\sigma$.

$\sigma \approx \dfrac{\$25,462 - \$17,627}{4} = 1958.75$

For insurer's cost:

$\dfrac{(n-1)s^2}{\chi^2_{\text{right}}} < \sigma^2 < \dfrac{(n-1)s^2}{\chi^2_{\text{left}}}$

$\dfrac{9(1958.75)^2}{21.666} < \sigma^2 < \dfrac{9(1958.75)^2}{2.088}$

$\$1,593,756 < \sigma^2 < \$16,537,507$

$\$1262.4 < \sigma < \$4066.6$

For uninsured person's cost:

$\sigma \approx \dfrac{\$58,702 - \$40,640}{4} = \$4515.5$

**7. continued**

$\dfrac{9(4515.5)^2}{21.666} < \sigma^2 < \dfrac{9(4515.5)^2}{2.088}$

$\$8,469,845 < \sigma^2 < \$87,886,811$

$\$2910.3 < \sigma < \$9374.8$

**9.**

$s = 36.56$

$\dfrac{(n-1)s^2}{\chi^2_{\text{right}}} < \sigma^2 < \dfrac{(n-1)s^2}{\chi^2_{\text{left}}}$

$\dfrac{5(36.56)^2}{11.071} < \sigma^2 < \dfrac{5(36.56)^2}{1.145}$

$604 < \sigma^2 < 5837$

$24.6 < \sigma < 76.4$

**11.**

$\dfrac{(n-1)s^2}{\chi^2_{\text{right}}} < \sigma^2 < \dfrac{(n-1)s^2}{\chi^2_{\text{left}}}$

$\dfrac{27(5.2)^2}{43.194} < \sigma^2 < \dfrac{27(5.2)^2}{14.573}$

$16.9 < \sigma^2 < 50.1$

$4.1 < \sigma < 7.1$

**13.**

$s - z_{\frac{\alpha}{2}}\left(\dfrac{s}{\sqrt{2n}}\right) < \sigma < s + z_{\frac{\alpha}{2}}\left(\dfrac{s}{\sqrt{2n}}\right)$

$18 - 1.96\left(\dfrac{18}{\sqrt{400}}\right) < \sigma < 18 + 1.96\left(\dfrac{18}{\sqrt{400}}\right)$

$16.2 < \sigma < 19.8$

## REVIEW EXERCISES - CHAPTER 7

**1.**

$\overline{X} = 19.63$ is the point estimate of $\mu$.

$\overline{X} - z_{\frac{\alpha}{2}}\left(\dfrac{s}{\sqrt{n}}\right) < \mu < \overline{X} + z_{\frac{\alpha}{2}}\left(\dfrac{s}{\sqrt{n}}\right)$

$19.63 - 1.65\left(\dfrac{18.73}{\sqrt{30}}\right) < \mu <$
$\qquad\qquad 19.63 + 1.65\left(\dfrac{18.73}{\sqrt{30}}\right)$

$19.63 - 5.64 < \mu < 19.63 + 5.64$

$13.99 < \mu < 25.27$ or $14 < \mu < 25$

(TI answer: $14.005 < \mu < 25.255$)

**3.**

$\overline{X} = 7.5$ is the point estimate of $\mu$.

$\overline{X} - z_{\frac{\alpha}{2}}\left(\dfrac{s}{\sqrt{n}}\right) < \mu < \overline{X} + z_{\frac{\alpha}{2}}\left(\dfrac{s}{\sqrt{n}}\right)$

$7.5 - 1.96\left(\dfrac{0.8}{\sqrt{1500}}\right) < \mu < 7.5 + 1.96\left(\dfrac{0.8}{\sqrt{1500}}\right)$

$7.46 < \mu < 7.54$

**5.**

$$\overline{X} - t_{\frac{\alpha}{2}}\left(\frac{s}{\sqrt{n}}\right) < \mu < \overline{X} + t_{\frac{\alpha}{2}}\left(\frac{s}{\sqrt{n}}\right)$$

$$28 - 2.132\left(\frac{3}{\sqrt{5}}\right) < \mu < 28 + 2.132\left(\frac{3}{\sqrt{5}}\right)$$

$$25 < \mu < 31$$

**7.**

$$n = \left[\frac{z_{\frac{\alpha}{2}}\sigma}{E}\right]^2 = \left[\frac{1.65(80)}{25}\right]^2$$

$$n = (5.28)^2 = 27.88 \text{ or } 28$$

**9.**

$$\hat{p} = 0.547 \qquad \hat{q} = 0.453$$

$$\hat{p} - (z_{\frac{\alpha}{2}})\sqrt{\frac{\hat{p}\hat{q}}{n}} < p < \hat{p} + (z_{\frac{\alpha}{2}})\sqrt{\frac{\hat{p}\hat{q}}{n}}$$

$$0.547 - 1.96\sqrt{\frac{(0.547)(0.453)}{75}} < p <$$
$$0.547 + 1.96\sqrt{\frac{(0.547)(0.453)}{75}}$$

$$0.547 - 0.113 < p < 0.547 + 0.113$$
$$0.434 < p < 0.660$$
Yes; it seems that as many as 66% were dissatisfied.

**11.**

$$\hat{p} = 0.88 \qquad \hat{q} = 0.12$$

$$n = \hat{p}\hat{q}\left[\frac{z_{\frac{\alpha}{2}}}{E}\right]^2 = (0.88)(0.12)\left[\frac{1.65}{0.025}\right]^2$$

$$n = 459.99 \text{ or } 460$$

**13.**

$$\frac{(n-1)s^2}{\chi^2_{right}} < \sigma^2 < \frac{(n-1)s^2}{\chi^2_{left}}$$

$$\frac{(18-1)(0.29)^2}{30.191} < \sigma^2 < \frac{(18-1)(0.29)^2}{7.564}$$

$$0.0474 < \sigma^2 < 0.1890$$
$$0.218 < \sigma < 0.435$$
Yes; it seems that there is a large standard deviation.

**15.**

$$\frac{(n-1)s^2}{\chi^2_{right}} < \sigma^2 < \frac{(n-1)s^2}{\chi^2_{left}}$$

$$\frac{(15-1)(8.6)}{23.685} < \sigma^2 < \frac{(15-1)(8.6)}{6.571}$$

$$5.1 < \sigma^2 < 18.3$$

**CHAPTER 7 QUIZ**

1. True
2. True
3. False, it is consistent if, as sample size increases, the estimator approaches the parameter being estimated.
4. True
5. b.
6. a.
7. b.
8. unbiased, consistent, relatively efficient
9. maximum error of estimate
10. point
11. 90, 95, 99

**12.**
$\overline{X} = \$23.45$ is the point estimate for $\mu$.
$$\overline{X} - z_{\frac{\alpha}{2}}\left(\frac{s}{\sqrt{n}}\right) < \mu < \overline{X} + z_{\frac{\alpha}{2}}\left(\frac{s}{\sqrt{n}}\right)$$

$$\$23.45 - 1.65\left(\frac{2.80}{\sqrt{49}}\right) < \mu <$$
$$\$23.45 + 1.65\left(\frac{2.80}{\sqrt{49}}\right)$$

$$\$22.79 < \mu < \$24.11$$

**13.**
$\overline{X} = \$44.80$ is the point estimate for $\mu$.

$$\overline{X} - t_{\frac{\alpha}{2}}\left(\frac{s}{\sqrt{n}}\right) < \mu < \overline{X} + t_{\frac{\alpha}{2}}\left(\frac{s}{\sqrt{n}}\right)$$

$$\$44.80 - 2.093\left(\frac{3.53}{\sqrt{20}}\right) < \mu <$$
$$\$44.80 + 2.093\left(\frac{3.53}{\sqrt{20}}\right)$$

$$\$43.15 < \mu < \$46.45$$

**14.**
$\overline{X} = 4150$ is the point estimate for $\mu$.

$$\overline{X} - z_{\frac{\alpha}{2}}\left(\frac{s}{\sqrt{n}}\right) < \mu < \overline{X} + z_{\frac{\alpha}{2}}\left(\frac{s}{\sqrt{n}}\right)$$

$$\$4150 - 2.58\left(\frac{480}{\sqrt{40}}\right) < \mu <$$
$$\$4150 + 2.58\left(\frac{480}{\sqrt{40}}\right)$$

$$\$3954 < \mu < \$4346$$

**15.**
$$\overline{X} - t_{\frac{\alpha}{2}}\left(\frac{s}{\sqrt{n}}\right) < \mu < \overline{X} + t_{\frac{\alpha}{2}}\left(\frac{s}{\sqrt{n}}\right)$$

$$48.6 - 2.262\left(\frac{4.1}{\sqrt{10}}\right) < \mu < 48.6 + 2.262\left(\frac{4.1}{\sqrt{10}}\right)$$

$$45.7 < \mu < 51.5$$

16.
$$\overline{X} - t_{\frac{\alpha}{2}}\left(\frac{s}{\sqrt{n}}\right) < \mu < \overline{X} + t_{\frac{\alpha}{2}}\left(\frac{s}{\sqrt{n}}\right)$$

$$438 - 3.499\left(\frac{16}{\sqrt{8}}\right) < \mu < 438 + 3.499\left(\frac{16}{\sqrt{8}}\right)$$

$$418 < \mu < 458$$

17.
$$\overline{X} - t_{\frac{\alpha}{2}}\left(\frac{s}{\sqrt{n}}\right) < \mu < \overline{X} + t_{\frac{\alpha}{2}}\left(\frac{s}{\sqrt{n}}\right)$$

$$31 - 2.353\left(\frac{4}{\sqrt{4}}\right) < \mu < 31 + 2.353\left(\frac{4}{\sqrt{4}}\right)$$

$$26 < \mu < 36$$

18.
$$n = \left[\frac{z_{\frac{\alpha}{2}}\,\sigma}{E}\right]^2 = \left[\frac{2.58(2.6)}{0.5}\right]^2$$

$$n = 179.98 \text{ or } 180$$

19.
$$n = \left[\frac{z_{\frac{\alpha}{2}}\,\sigma}{E}\right]^2 = \left[\frac{1.65(900)}{300}\right]^2$$

$$n = 24.5 \text{ or } 25$$

20.
$$\hat{p} - (z_{\frac{\alpha}{2}})\sqrt{\frac{\hat{p}\hat{q}}{n}} < p < \hat{p} + (z_{\frac{\alpha}{2}})\sqrt{\frac{\hat{p}\hat{q}}{n}}$$

$$\hat{p} = \frac{53}{75} = 0.707 \quad \hat{q} = \frac{22}{75} = 0.293$$

$$0.71 - 1.96\sqrt{\frac{(0.707)(0.293)}{75}} < p <$$
$$0.71 + 1.96\sqrt{\frac{(0.707)(0.293)}{75}}$$

$$0.604 < p < 0.810$$

21.
$$\hat{p} - (z_{\frac{\alpha}{2}})\sqrt{\frac{\hat{p}\hat{q}}{n}} < p < \hat{p} + (z_{\frac{\alpha}{2}})\sqrt{\frac{\hat{p}\hat{q}}{n}}$$

$$0.36 - 1.65\sqrt{\frac{(0.36)(0.64)}{150}} < p <$$
$$0.36 + 1.65\sqrt{\frac{(0.36)(0.64)}{150}}$$

$$0.295 < p < 0.425$$

22.
$$\hat{p} - (z_{\frac{\alpha}{2}})\sqrt{\frac{\hat{p}\hat{q}}{n}} < p < \hat{p} + (z_{\frac{\alpha}{2}})\sqrt{\frac{\hat{p}\hat{q}}{n}}$$

22. continued
$$0.4444 - 1.96\sqrt{\frac{(0.4444)(0.5556)}{90}} < p <$$
$$0.4444 + 1.96\sqrt{\frac{(0.4444)(0.5556)}{90}}$$

$$0.342 < p < 0.547$$

23.
$$n = \hat{p}\,\hat{q}\left[\frac{z_{\frac{\alpha}{2}}}{E}\right]^2$$

$$n = (0.15)(0.85)\left[\frac{1.96}{0.03}\right]^2$$

$$n = 544.22 \text{ or } 545$$

24.
$$\frac{(n-1)s^2}{\chi^2_{\text{right}}} < \sigma^2 < \frac{(n-1)s^2}{\chi^2_{\text{left}}}$$

$$\frac{24(9)^2}{39.364} < \sigma^2 < \frac{24(9)^2}{12.401}$$

$$49.4 < \sigma^2 < 156.8$$
$$7 < \sigma < 13$$

25.
$$\frac{(n-1)s^2}{\chi^2_{\text{right}}} < \sigma^2 < \frac{(n-1)s^2}{\chi^2_{\text{left}}}$$

$$\frac{26(6.8)^2}{38.885} < \sigma^2 < \frac{26(6.8)^2}{15.379}$$

$$30.9 < \sigma^2 < 78.2$$
$$5.6 < \sigma < 8.8$$

26.
$$\frac{(n-1)s^2}{\chi^2_{\text{right}}} < \sigma^2 < \frac{(n-1)s^2}{\chi^2_{\text{left}}}$$

$$\frac{19(2.3)^2}{30.144} < \sigma^2 < \frac{19(2.3)^2}{10.177}$$

$$3.33 < \sigma^2 < 10$$
$$1.8 < \sigma < 3.2$$

Note: Graphs are not to scale and are intended to convey a general idea. Answers may vary due to rounding.

EXERCISE SET 8-1

1.
The null hypothesis is a statistical hypothesis that states there is no difference between a parameter and a specific value or there is no difference between two parameters. The alternative hypothesis specifies a specific difference between a parameter and a specific value, or that there is a difference between two parameters. Examples will vary.

3.
A statistical test uses the data obtained from a sample to make a decision as to whether or not the null hypothesis should be rejected.

5.
The critical region is the region of values of the test-statistic that indicates a significant difference and the null hypothesis should be rejected. The non-critical region is the region of values of the test-statistic that indicates the difference was probably due to chance, and the null hypothesis should not be rejected.

7.
Type I is represented by $\alpha$, type II is represented by $\beta$.

9.
A one-tailed test should be used when a specific direction, such as greater than or less than, is being hypothesized, whereas when no direction is specified, a two-tailed test should be used.

11.
Hypotheses can only be proved true when the entire population is used to compute the test statistic. In most cases, this is impossible.

12.
a. $\pm 1.96$

12a. continued

− 1.96    0    + 1.96

b. − 2.33

− 2.33    0

c. + 2.58

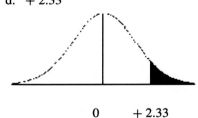

0    + 2.58

d. + 2.33

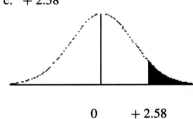

0    + 2.33

e. − 1.65

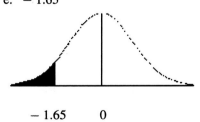

− 1.65    0

12. continued

f.  $-2.05$

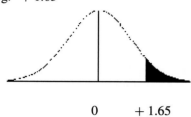

$-2.05 \quad 0$

g.  $+1.65$

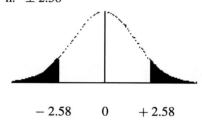

$0 \quad +1.65$

h.  $\pm 2.58$

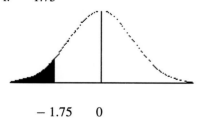

$-2.58 \quad 0 \quad +2.58$

i.  $-1.75$

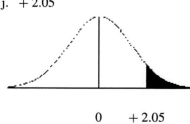

$-1.75 \quad 0$

j.  $+2.05$

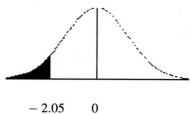

$0 \quad +2.05$

13.

a.  $H_0$: $\mu = 24.6$
    $H_1$: $\mu \neq 24.6$

b.  $H_0$: $\mu = \$51{,}497$
    $H_1$: $\mu \neq \$51{,}497$

13. continued

c.  $H_0$: $\mu = 25.4$
    $H_1$: $\mu > 25.4$

d.  $H_0$: $\mu = 88$
    $H_1$: $\mu < 88$

e.  $H_0$: $\mu = 70$
    $H_1$: $\mu < 70$

f.  $H_0$: $\mu = \$79.95$
    $H_1$: $\mu \neq \$79.95$

g.  $H_0$: $\mu = 8.2$
    $H_1$: $\mu \neq 8.2$

**EXERCISE SET 8-2**

1.
$H_0$: $\mu = 5000$
$H_1$: $\mu > 5000$ (claim)

C. V. $= 1.65$
$z = \dfrac{\overline{X} - \mu}{\frac{\sigma}{\sqrt{n}}} = \dfrac{5430 - 5000}{\frac{600}{\sqrt{40}}} = 4.53$

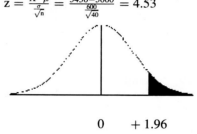

$0 \quad +1.96 \quad \underset{4.53}{\uparrow}$

Reject the null hypothesis. There is enough evidence at $\alpha = 0.05$ to conclude that the mean is greater than 5000 steps.

3.
$H_0$: $\mu = \$24$ billion
H1: $\mu > \$24$ billion  (claim)

C. V. $= +1.65$ $\quad \overline{X} = \$31.5 \quad s = \$28.7$
$z = \dfrac{\overline{X} - \mu}{\frac{\sigma}{\sqrt{n}}} = \dfrac{31.5 - 24}{\frac{28.7}{\sqrt{50}}} = 1.85$

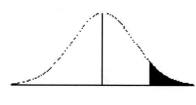

$0 \quad 1.65 \quad \underset{1.85}{\uparrow}$

3. continued
Reject the null hypothesis. There is enough
evidence to support the claim that the
average revenue exceeds $24 billion.

5.
$H_0$: $\mu = 1468$
$H_1$: $\mu \neq 1468$ (claim)

C. V. $= \pm 2.58$ (0.01); $\pm 1.96$ (0.05)

$z = \frac{\overline{X} - \mu}{\frac{s}{\sqrt{n}}} = \frac{\$1520 - \$1468}{\frac{198}{\sqrt{60}}} = 2.03$

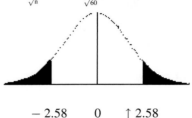

$-2.58 \quad 0 \quad \uparrow 2.58$
$2.03$

Do not reject at 0.01. There is not enough
evidence to support the claim that average
expenditure differs from $1468. Reject at
0.05.

7.
$H_0$: $\mu = 29$
$H_1$: $\mu \neq 29$ (claim)

C. V. $= \pm 1.96$ $\quad \overline{X} = 29.45 \quad s = 2.61$

$z = \frac{\overline{X} - \mu}{\frac{\sigma}{\sqrt{n}}} = \frac{29.45 - 29}{\frac{2.61}{\sqrt{30}}} = 0.944$

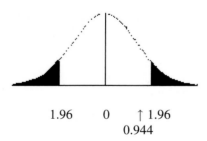

$1.96 \quad 0 \quad \uparrow 1.96$
$0.944$

Do not reject the null hypothesis. There is
enough evidence to reject the claim that the
average height differs from 29 inches.

9.
$H_0$: $\mu = \$26,025$
$H_1$: $\mu > \$26,025$ (claim)

C. V. $= 1.65$

$z = \frac{\overline{X} - \mu}{\frac{\sigma}{\sqrt{n}}} = \frac{\$27,690 - \$26,025}{\frac{5492}{\sqrt{40}}} = 1.92$

9. continued

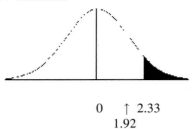

$0 \quad \uparrow 2.33$
$1.92$

Reject the null hypothesis. There is enough
evidence to support the claim that the
average tuition cost has increased.

11.
$H_0$: $\mu = 500$
$H_1$: $\mu \neq 500$ (claim)

C. V. $= \pm 2.58$

$z = \frac{\overline{X} - \mu}{\frac{\sigma}{\sqrt{n}}} = \frac{476 - 500}{\frac{42}{\sqrt{50}}} = -4.04$

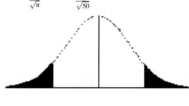

$\uparrow \quad -2.58 \quad 0 \quad 2.58$
$-4.04$

Reject the null hypothesis. There is enough
evidence to conclude that the average
number of cups differs from 500.

13.
$H_0$: $\mu = 60.35$
$H_1$: $\mu < 60.35$ (claim)

C. V. $= -1.65$

$z = \frac{\overline{X} - \mu}{\frac{s}{\sqrt{n}}} = \frac{55.4 - 60.35}{\frac{6.5}{\sqrt{40}}} = -4.82$

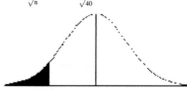

$\uparrow \quad -1.65 \quad 0$
$-4.82$

Reject the null hypothesis. There is enough
evidence to support the claim that state
senators are younger than U. S. Senators.

15.
a. Do not reject.
b. Reject.
c. Do not reject.
d. Reject
e. Reject

17.
$H_0$: $\mu = 264$
$H_1$: $\mu < 264$    (claim)
$z = \dfrac{X - \mu}{\frac{\sigma}{\sqrt{n}}} = \dfrac{262.3 - 264}{\frac{3}{\sqrt{20}}} = -2.53$

The area corresponding to $z = 2.53$ is
0.9943. The P-value is $1 - 0.9943 =$
0.0057. The decision is to reject the null
hypothesis since $0.0057 < 0.01$. There is
enough evidence to support the claim that
the average stopping distance is less than
264 feet.
(TI: P-value = 0.0056)

19.
$H_0$: $\mu = 546$
$H_1$: $\mu < 546$    (claim)
$z = \dfrac{X - \mu}{\frac{\sigma}{\sqrt{n}}} = \dfrac{544.8 - 546}{\frac{3}{\sqrt{36}}} = -2.4$

The area corresponding to $z = -2.4$ is
0.0082. Thus, P-value = 0.0082. The
decision is to reject the null hypothesis since
$0.0082 < 0.01$. There is enough evidence to
support the claim that the number of calories
burned is less than 546.
(TI: P-value = 0.0082)

21.
$H_0$: $\mu = 444$
$H_1$: $\mu \neq 444$
$z = \dfrac{X - \mu}{\frac{\sigma}{\sqrt{n}}} = \dfrac{430 - 444}{\frac{52}{\sqrt{40}}} = -1.70$

The area corresponding to $z = -1.70$ is
0.0446. To get the P-value, multiply by 2
since this is a two-tailed test. Thus,
P-value $= 2(0.0446) = 0.0892$.
The decision is do not reject the null
hypothesis since P-value $> 0.05$. There is
not enough evidence to support the claim
that the mean differs from 444.
(TI: P-value = 0.0886)

23.
$H_0$: $\mu = 30{,}000$    (claim)
$H_1$: $\mu \neq 30{,}000$

23. continued
$z = \dfrac{X - \mu}{\frac{s}{\sqrt{n}}} = \dfrac{30{,}456 - 30{,}000}{\frac{1684}{\sqrt{40}}} = 1.71$

The area corresponding to $z = 1.71$ is
0.9564. The P-value is $2(1 - 0.9564) =$
$2(0.0436) = 0.0872$. The decision is to
reject the null hypothesis at $\alpha = 0.10$ since
$0.0872 < 0.10$. The conclusion is that there
is enough evidence to reject the claim that
customers are adhering to the
recommendation. A 0.10 significance level
is probably appropriate since there is little
consequence of a Type I error. The dealer
would be advised to increase efforts to make
its customers aware of the service
recommendation.
(TI: P-value = 0.0868)

25.
$H_0$: $\mu = 10$
$H_1$: $\mu < 10$    (claim)

$X = 5.025$    $s = 3.63$
$z = \dfrac{X - \mu}{\frac{s}{\sqrt{n}}} = \dfrac{5.025 - 10}{\frac{3.63}{\sqrt{40}}} = -8.67$

The area corresponding to 8.67 is greater
than 0.9999. The P-value is
$1 - 0.9999 < 0.0001$. Since $0.0001 < 0.05$,
the decision is to reject the null hypothesis.
There is enough evidence to support the
claim that the average number of days
missed per year is less than 10.

27.
The mean and standard deviation are found
as follows:

|           | f  | $X_m$ | $f \cdot X_m$ | $f \cdot X_m^2$ |
|-----------|----|-------|---------------|-----------------|
| 8.35 - 8.43 | 2  | 8.39  | 16.78         | 140.7842        |
| 8.44 - 8.52 | 6  | 8.48  | 50.88         | 431.4624        |
| 8.53 - 8.61 | 12 | 8.57  | 102.84        | 881.3388        |
| 8.62 - 8.70 | 18 | 8.66  | 155.88        | 1349.9208       |
| 8.71 - 8.79 | 10 | 8.75  | 87.5          | 765.625         |
| 8.80 - 8.88 | 2  | 8.84  | 17.68         | 156.2912        |
|           | 50 |       | 431.56        | 3725.4224       |

$\overline{X} = \dfrac{\sum f \cdot X_m}{n} = \dfrac{431.56}{50} = 8.63$

$s = \sqrt{\dfrac{\sum f \cdot X_m^2 - \frac{\left(\sum f \cdot X_m\right)^2}{n}}{n - 1}} = \sqrt{\dfrac{3725.4224 - \frac{(431.56)^2}{50}}{49}}$

$s = 0.105$

**27. continued**

$H_0$: $\mu = 8.65$    (claim)

$H_1$: $\mu \neq 8.65$

C. V. $= \pm 1.96$

$z = \dfrac{\overline{X}-\mu}{\frac{s}{\sqrt{n}}} = \dfrac{8.63-8.65}{\frac{0.105}{\sqrt{50}}} = -1.35$

Do not reject the null hypothesis. There is not enough evidence to reject the claim that the average hourly wage of the employees is $8.65.

**EXERCISE SET 8-3**

**1.**

It is bell-shaped, symmetric about the mean, and it never touches the x axis. The mean, median, and mode are all equal to 0 and they are located at the center of the distribution. The t distribution differs from the standard normal distribution in that it is a family of curves, the variance is greater than one, and as the degrees of freedom increase the t distribution approaches the standard normal distribution.

**3.**

a. d. f. $= 9$      C. V. $= +1.833$

b. d. f. $= 17$     C. V. $= \pm 1.740$

c. d. f. $= 5$      C. V. $= -3.365$

d. d. f. $= 8$      C. V. $= +2.306$

e. d. f. $= 14$     C. V. $= \pm 2.145$

f. d. f. $= 22$     C. V. $= -2.819$

g. d. f. $= 27$     C. V. $= \pm 2.771$

h. d. f. $= 16$     C. V. $= \pm 2.583$

**4.**

a. $0.01 <$ P-value $< 0.025$   (0.018)

b. $0.05 <$ P-value $< 0.10$   (0.062)

c. $0.10 <$ P-value $< 0.25$   (0.123)

d. $0.10 <$ P-value $< 0.20$   (0.138)

e. P-value $< 0.005$        (0.003)

f. $0.10 <$ P-value $< 0.25$   (0.158)

g. P-value $= 0.05$        (0.05)

h. P-value $> 0.25$        (0.261)

**5.**

$H_0$: $\mu = 179$

$H_1$: $\mu \neq 179$    (claim)

C. V. $= \pm 3.250$    d. f. $= 9$

$t = \dfrac{\overline{X}-\mu}{\frac{s}{\sqrt{n}}} = \dfrac{205-179}{\frac{26}{\sqrt{10}}} = 3.162$

**5. continued**

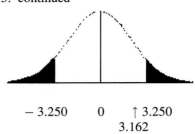

$-3.250$     $0$     $\uparrow$ 3.250

                   3.162

Do not reject the null hypothesis. There is not enough evidence to support the claim that the average expense is $179.

**7.**

$H_0$: $\mu = \$4172$

$H_1$: $\mu > \$4172$    (claim)

C. V. $= 1.729$    d. f. $= 19$

$t = \dfrac{\overline{X}-\mu}{\frac{s}{\sqrt{n}}} = \dfrac{4560-4172}{\frac{1590}{\sqrt{20}}} = 1.091$

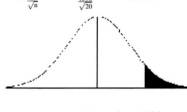

      $0$     $\uparrow$   1.729

              1.091

Do not reject the null hypothesis. There is not enough evidence to support the claim that the average tax is more than $4172.

**9.**

$H_0$: $\mu = 700$    (claim)

$H_1$: $\mu < 700$

$\overline{X} = 606.5$    $s = 109.1$

C. V. $= -2.262$    d. f. $= 9$

$t = \dfrac{\overline{X}-\mu}{\frac{s}{\sqrt{n}}} = \dfrac{606.5-700}{\frac{109.1}{\sqrt{10}}} = -2.71$

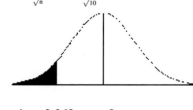

$\uparrow$ $-2.262$     $0$

$-2.71$

Reject the null hypothesis. There is enough evidence to reject the claim that the average height of the buildings is at least 700 feet.

**11.**
$H_0$: $\mu = \$13,252$
$H_1$: $\mu > \$13,252$   (claim)

C. V. = 2.539        d. f. = 19
$t = \frac{\overline{X} - \mu}{\frac{s}{\sqrt{n}}} = \frac{\$15,560 - \$13,252}{\frac{\$3500}{\sqrt{19}}} = 2.949$

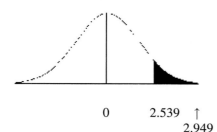

0        2.539  ↑
                2.949

Reject the null hypothesis. There is enough evidence to support the claim that the average tuition cost has increased.

**13.**
$H_0$: $\mu = \$54.8$
$H_1$: $\mu > \$54.8$   (claim)

C. V. = 1.761        d. f. = 14
$t = \frac{\overline{X} - \mu}{\frac{s}{\sqrt{n}}} = \frac{\$62.3 - \$54.8}{\frac{\$9.5}{\sqrt{15}}} = 3.058$

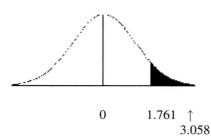

0        1.761  ↑
                3.058

Reject the null hypothesis. There is enough evidence to support the claim that the cost to produce an action movie is more than $54.8 million.

**15.**
$H_0$: $\mu = \$623$
$H_1$: $\mu > \$623$   (claim)

C. V.  = 2.528    d. f. = 20
$t = \frac{\overline{X} - \mu}{\frac{s}{\sqrt{n}}} = \frac{650 - 623}{\frac{72}{\sqrt{21}}} = 1.718$

**15. continued**

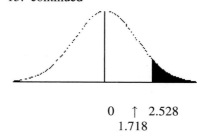

0   ↑  2.528
     1.718

Do not reject the null hypothesis. There is not enough evidence to support the claim that the average weekly earnings are more than $623.

**17.**
$H_0$: $\mu = 5.8$
$H_1$: $\mu \neq 5.8$   (claim)
$\overline{X} = 3.85$     s = 2.52
d. f. = 19      $\alpha = 0.05$
P-value < 0.01     (0.0026)
$t = \frac{\overline{X} - \mu}{\frac{s}{\sqrt{n}}} = \frac{3.85 - 5.8}{\frac{2.52}{\sqrt{20}}} = -3.46$

Since P-value < 0.01, reject the null hypothesis. There is enough evidence to support the claim that the mean is not 5.8.

**19.**
$H_0$: $\mu = \$15,000$
$H_1$: $\mu \neq \$15,000$   (claim)

$\overline{X} = \$14,347.17$     s = \$2048.54
d. f.  = 11    C. V. = $\pm 2.201$

$t = \frac{\overline{X} - \mu}{\frac{s}{\sqrt{n}}} = \frac{\$14,347.17 - \$15,000}{\frac{\$2048.54}{\sqrt{12}}} = -1.10$

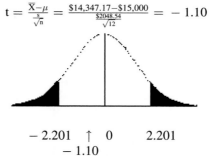

$-2.201$  ↑  0        2.201
      $-1.10$

Do not reject the null hypothesis. There is not enough evidence to say that the average stipend differs from $15,000.

**EXERCISE SET 8-4**

**1.**
Answers will vary.

3.
$np \geq 5$ and $nq \geq 5$

5.
$H_0$: $p = 0.686$
$H_1$: $p \neq 0.686$   (claim)

$\hat{p} = \frac{92}{150} = 0.613$    $p = 0.686$    $q = 0.314$
C. V. $= \pm 2.58$

$z = \frac{\hat{p}-p}{\sqrt{\frac{pq}{n}}} = \frac{0.613-0.686}{\sqrt{\frac{(0.686)(0.314)}{150}}} = -1.93$
(TI: $z = -1.918$)

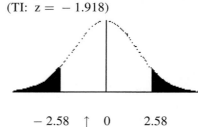

$-2.58$   $\uparrow$   $0$     $2.58$
      $-1.93$

Do not reject the null hypothesis. There is not enough evidence to support the claim that the proportion of homeowners is different from 68.6%.

7.
$H_0$: $p = 0.40$
$H_1$: $p \neq 0.40$   (claim)

$\hat{p} = \frac{65}{180} = 0.361$    $p = 0.40$    $q = 0.60$
C. V. $= \pm 2.58$
$z = \frac{\hat{p}-p}{\sqrt{\frac{pq}{n}}} = \frac{0.361-0.40}{\sqrt{\frac{(0.40)(0.60)}{180}}} = -1.07$

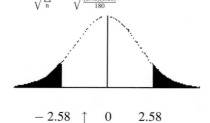

$-2.58$   $\uparrow$   $0$     $2.58$
      $-1.07$

Do not reject the null hypothesis. There is not enough evidence to conclude that the proportion differs from 40%.

9.
$H_0$: $p = 0.78$   (claim)
$H_1$: $p \neq 0.78$

$\hat{p} = \frac{100}{143} = 0.6993$    $p = 0.78$    $q = 0.22$
C. V. $= \pm 1.96$

9. continued
$z = \frac{\hat{p}-p}{\sqrt{\frac{pq}{n}}} = \frac{0.6993-0.78}{\sqrt{\frac{(0.78)(0.27)}{143}}} = -2.33$

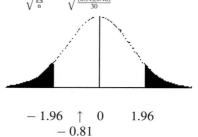

$\uparrow$   $-1.96$     $0$     $1.96$
$-2.33$

Reject the null hypothesis. There is enough evidence to reject the claim that the percentage is 78%.

11.
$H_0$: $p = 0.54$
$H_1$: $p \neq 0.54$   (claim)

$\hat{p} = \frac{14}{30} = 0.4667$    $p = 0.54$    $q = 0.46$
C. V. $= \pm 1.96$
$z = \frac{\hat{p}-p}{\sqrt{\frac{pq}{n}}} = \frac{0.4667-0.54}{\sqrt{\frac{(0.54)(0.46)}{30}}} = -0.81$

$-1.96$   $\uparrow$   $0$     $1.96$
     $-0.81$

Do not reject the null hypothesis. There is not enough evidence to reject the claim that 54% of fatal car/truck accidents are caused by driver error.

13.
$H_0$: $p = 0.54$   (claim)
$H_1$: $p \neq 0.54$

$\hat{p} = \frac{36}{60} = 0.6$    $p = 0.54$    $q = 0.46$
$z = \frac{\hat{p}-p}{\sqrt{\frac{pq}{n}}} = \frac{0.6-0.54}{\sqrt{\frac{(0.54)(0.46)}{60}}} = 0.93$
Area $= 0.8238$
P-value $= 2(1 - 0.8238) = 0.3524$
Since P-value $> 0.01$, do not reject the null hypothesis. There is enough evidence to support the claim that 54% of kids had a snack after school. Yes, a healthy snack should be made available for children to eat after school.
(TI: P-value $= 0.3511$)

**15.**
$H_0$: $p = 0.18$    (claim)
$H_1$: $p \neq 0.18$

$\hat{p} = \frac{50}{300} = 0.1667$    $p = 0.18$    $q = 0.82$
$z = \frac{\hat{p} - p}{\sqrt{\frac{pq}{n}}} = \frac{0.1667 - 0.18}{\sqrt{\frac{(0.18)(0.82)}{300}}} = -0.60$
Area $= 0.2743$
P-value $= 2(0.2743) = 0.5486$
Since P-value $> 0.05$, do not reject the null hypothesis. There is not enough evidence to reject the claim that 18% of all high school students smoke at least a pack of cigarettes a day. (TI: P-value $= 0.5478$)

**17.**
$H_0$: $p = 0.67$
$H_1$: $p \neq 0.67$    (claim)

$\hat{p} = \frac{82}{100} = 0.82$    $p = 0.67$    $q = 0.33$
C. V. $= \pm 1.96$
$z = \frac{\hat{p} - p}{\sqrt{\frac{pq}{n}}} = \frac{0.82 - 0.67}{\sqrt{\frac{(0.67)(0.33)}{100}}} = 3.19$

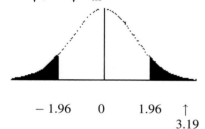

$-1.96$    $0$    $1.96$    $\uparrow$
$3.19$

Reject the null hypothesis. There is enough evidence to support the claim that the percentage is not 67%.

**19.**
$H_0$: $p = 0.576$
$H_1$: $p < 0.576$   (claim)

$\hat{p} = \frac{17}{36} = 0.472$    $p = 0.576$    $q = 0.424$
C. V. $= -1.65$
$z = \frac{\hat{p} - p}{\sqrt{\frac{pq}{n}}} = \frac{0.472 - 0.576}{\sqrt{\frac{(0.576)(0.424)}{36}}} = -1.26$

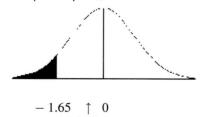

$-1.65$    $\uparrow$    $0$
$-1.26$

**19. continued**
Do not reject the null hypothesis. There is not enough evidence to support the claim that the percentage of injuries during practice is below 57.6%.

**21.**
This represents a binomial distribution with $p = 0.50$ and $n = 9$. The P-value is $2 \cdot P(X \leq 3) = 2(0.254) = 0.508$.

Since P-value $> 0.10$, the conclusion that the coin is not balanced is probably false.

**23.**
$z = \frac{X - \mu}{\sigma}$

$z = \frac{X - np}{\sqrt{npq}}$

$z = \frac{\frac{X}{n} - \frac{np}{n}}{\frac{1}{n}\sqrt{npq}}$

$z = \frac{\frac{X}{n} - \frac{np}{n}}{\sqrt{\frac{npq}{n^2}}}$

$z = \frac{\hat{p} - p}{\sqrt{\frac{pq}{n}}}$

**EXERCISE SET 8-5**

**1.**
a.  $H_0$: $\sigma^2 = 225$
    $H_1$: $\sigma^2 > 225$

C. V. $= 27.587$    d. f. $= 17$

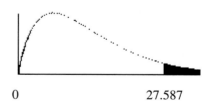

$0$                    $27.587$

b.  $H_0$: $\sigma^2 = 225$
    $H_1$: $\sigma^2 < 225$

C. V. $= 14.042$    d. f. $= 22$

1b. continued

0   14.042

c. $H_0$: $\sigma^2 = 225$
   $H_1$: $\sigma^2 \neq 225$

C. V. = 5.629, 26.119    d. f. = 14

0   5.629                26.119

d. $H_0$: $\sigma^2 = 225$
   $H_1$: $\sigma^2 \neq 225$

C. V. = 2.167, 14.067    d. f. = 7

0   2.167              14.067

e. $H_0$: $\sigma^2 = 225$
   $H_1$: $\sigma^2 > 225$

C. V. = 32.000    d. f. = 16

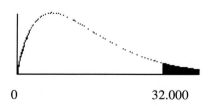

0                        32.000

f. $H_0$: $\sigma^2 = 225$
   $H_1$: $\sigma^2 < 225$

C. V. = 8.907    d. f. = 19

1f. continued

0   8.907

g. $H_0$: $\sigma^2 = 225$
   $H_1$: $\sigma^2 \neq 225$

C. V. = 3.074, 28.299    d. f. = 12

0   3.074                28.299

h. $H_0$: $\sigma^2 = 225$
   $H_1$: $\sigma^2 < 225$

C. V. = 15.308    d. f. = 28

0   15.308

2.
a.  0.01 < P-value < 0.025   (0.015)
b.  0.005 < P-value < 0.01   (0.006)
c.  0.01 < P-value < 0.025   (0.012)
d.  P-value < 0.005            (0.003)
e.  0.025 < P-value < 0.05   (0.037)
f.  0.05 < P-value < 0.10    (0.088)
g.  0.05 < P-value < 0.10    (0.066)
h.  P-value < 0.01            (0.007)

3.
$H_0$: $\sigma = 60$        (claim)
$H_1$: $\sigma \neq 60$

C. V. = 8.672, 27.587    $\alpha = 0.10$
d. f. = 17
s = 64.6
$\chi^2 = \frac{(n-1)s^2}{\sigma^2} = \frac{(18-1)(64.6)^2}{(60)^2} = 19.707$

3. continued

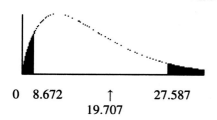

0   8.672          ↑          27.587
                 19.707

Do not reject the null hypothesis. There is not enough evidence to reject the claim that the standard deviation is 60.

5.
$H_0$: $\sigma = 15$
$H_1$: $\sigma < 15$   (claim)

C. V. = 4.575   $\alpha = 0.05$   d. f. = 14

$\chi^2 = \frac{(n-1)s^2}{\sigma^2} = \frac{(15-1)(13.6)^2}{15^2} = 9.0425$

0  4.575    ↑
           9.0425

Do not reject the null hypothesis. There is not enough evidence to support the claim that the standard deviation is less than 15.

7.
$H_0$: $\sigma = 1.2$   (claim)
$H_1$: $\sigma > 1.2$

$\alpha = 0.01$   d. f. = 14

$\chi^2 = \frac{(n-1)s^2}{\sigma^2} = \frac{(15-1)(1.8)^2}{(1.2)^2} = 31.5$

P-value < 0.005   (0.0047)
Since P-value < 0.01, reject the null hypothesis. There is enough evidence to reject the claim that the standard deviation is less than or equal to 1.2 minutes.

9.
$H_0$: $\sigma = 20$
$H_1$: $\sigma > 20$   (claim)

s = 35.11

9. continued
C. V. = 36.191   $\alpha = 0.01$   d. f. = 19

$\chi^2 = \frac{(n-1)s^2}{\sigma^2} = \frac{(20-1)(35.11)^2}{20^2} = 58.55$

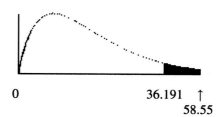

0                          36.191  ↑
                                  58.55

Reject the null hypothesis. There is enough evidence to support the claim that the standard deviation is more than 20 calories.

11.
$H_0$: $\sigma = 35$
$H_1$: $\sigma < 35$   (claim)

C. V. = 3.940   $\alpha = 0.05$   d. f. = 10

$\chi^2 = \frac{(n-1)s^2}{\sigma^2} = \frac{(11-1)(32)^2}{35^2} = 8.3592$

0   3.940      ↑
             8.3592

Do not reject the null hypothesis. There is not enough evidence to support the claim that the standard deviation is less than 35.

13.
$H_0$: $\sigma = 25$
$H_1$: $\sigma > 25$   (claim)

C. V. = 22.362   $\alpha = 0.05$   d. f. = 13

$\chi^2 = \frac{(n-1)s^2}{\sigma^2} = \frac{(14-1)(6.74)^2}{25} = 23.622$

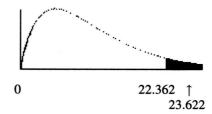

0                          22.362  ↑
                                  23.622

13. continued
Reject the null hypothesis. There is enough evidence to support the claim that the variance is greater than 25.

15.
$H_0$: $\sigma = 0.52$
$H_1$: $\sigma > 0.52$    (claim)

C. V. $= 30.144$    $\alpha = 0.05$    d. f. $= 19$

$$\chi^2 = \frac{(n-1)s^2}{\sigma^2} = \frac{(20-1)(0.568)^2}{(0.52)^2} = 22.670$$

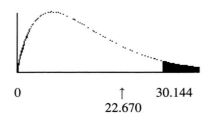

0          $\uparrow$    30.144
          22.670

Do not reject the null hypothesis. There is not enough evidence to support the claim that the standard deviation is more than 0.52 mm.

EXERCISE SET 8-6

1.
$H_0$: $\mu = 1800$    (claim)
$H_1$: $\mu \neq 1800$

C. V. $= \pm 1.96$
$$z = \frac{\overline{X}-\mu}{\frac{\sigma}{\sqrt{n}}} = \frac{1830-1800}{\frac{200}{\sqrt{10}}} = 0.47$$

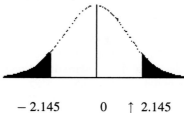

$-1.96$       0 $\uparrow$    1.96
           0.47

The 95% confidence interval of the mean is:
$$\overline{X} - z_{\frac{\alpha}{2}}\frac{\sigma}{\sqrt{n}} < \mu < \overline{X} + z_{\frac{\alpha}{2}}\frac{\sigma}{\sqrt{n}}$$

$1830 - 1.96\left(\frac{200}{\sqrt{10}}\right) < \mu <$
           $1830 + 1.96\left(\frac{200}{\sqrt{10}}\right)$

$1706.04 < \mu < 1953.96$

1. continued
The hypothesized mean is within the interval, thus we can be 95% confident that the average sales will be between $1706.94 and $1953.96.

3.
$H_0$: $\mu = 86$    (claim)
$H_1$: $\mu \neq 86$

C. V. $= \pm 2.58$
$$z = \frac{\overline{X}-\mu}{\frac{\sigma}{\sqrt{n}}} = \frac{84-86}{\frac{6}{\sqrt{15}}} = -1.29$$

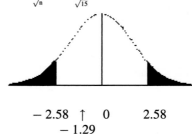

$-2.58$ $\uparrow$   0      2.58
     $-1.29$

$$\overline{X} - z_{\frac{\alpha}{2}}\frac{\sigma}{\sqrt{n}} < \mu < \overline{X} + z_{\frac{\alpha}{2}}\frac{\sigma}{\sqrt{n}}$$

$$84 - 2.58 \cdot \frac{6}{\sqrt{15}} < \mu < 84 + 1.58 \cdot \frac{6}{\sqrt{15}}$$

$80.00 < \mu < 88.00$

The decision is do not reject the null hypothesis since $-1.29 > -2.58$ and the 99% confidence interval contains the hypothesized mean. There is not enough evidence to reject the claim that the monthly maintenance is $86.

5.
$H_0$: $\mu = 19$
$H_1$: $\mu \neq 19$    (claim)

C. V. $= \pm 2.145$
$$t = \frac{\overline{X}-\mu}{\frac{s}{\sqrt{n}}} = \frac{21.3-19}{\frac{6.5}{\sqrt{15}}} = 1.37$$

$-2.145$       0    $\uparrow$ 2.145
          1.37

The 99% confidence interval of the mean is:

**5. continued**

$$\overline{X} - z_{\frac{\alpha}{2}} \frac{\sigma}{\sqrt{n}} < \mu < \overline{X} + z_{\frac{\alpha}{2}} \frac{\sigma}{\sqrt{n}}$$

$$21.3 - 2.145 \cdot \frac{6.5}{\sqrt{15}} < \mu < 21.3 + 2.145 \cdot \frac{6.5}{\sqrt{15}}$$

$$17.7 < \mu < 24.9$$

The decision is do not reject the null hypothesis since $1.37 < 2.145$ and the 99% confidence interval does contain the hypothesized mean of 19. The conclusion is that there is not enough evidence to support the claim that the average time worked at home is not 19 hours per week.

**7.**
The power of a statistical test is the probability of rejecting the null hypothesis when it is false.

**9.**
The power of a test can be increased by increasing $\alpha$ or selecting a larger sample size.

**REVIEW EXERCISES - CHAPTER 8**

**1.**
$H_0$: $\mu = 98°$    (claim)
$H_1$: $\mu \neq 98°$

C. V. $= \pm 1.96$
$$z = \frac{\overline{X} - \mu}{\frac{s}{\sqrt{n}}} = \frac{95.8 - 98}{\frac{7.71}{\sqrt{50}}} = -2.02$$

$\uparrow -1.96 \qquad 0 \qquad 1.96$
$-2.02$

Reject the null hypothesis. There is enough evidence to reject the claim that the average high temperature is 98°.

**3.**
$H_0$: $\mu = \$1229$
$H_1$: $\mu \neq \$1229$    (claim)

C. V. $= \pm 1.96$

**3. continued**
$$t = \frac{\overline{X} - \mu}{\frac{s}{\sqrt{n}}} = \frac{\$1350 - \$1229}{\frac{\$250}{\sqrt{15}}} = 1.875$$

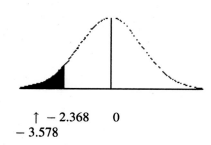

$-1.96 \qquad 0 \qquad \uparrow 1.96$
$\qquad\qquad\qquad 1.875$

Do not reject the null hypothesis. There is not enough evidence to support the claim that the average rent differs from $1229.

**5.**
$H_0$: $\mu = 18{,}000$
$H_1$: $\mu < 18{,}000$    (claim)

$\overline{X} = 16{,}298.37$    $s = 2604.82$
C. V. $= -2.368$

$$z = \frac{\overline{X} - \mu}{\frac{s}{\sqrt{n}}} = \frac{16{,}298.37 - 18{,}000}{\frac{2604.82}{\sqrt{30}}} = -3.578$$

$\uparrow -2.368 \qquad 0$
$-3.578$

Reject the null hypothesis. There is enough evidence to support the claim that average debt is less than $18,000.

**7.**
$H_0$: $\mu = 208$
$H_1$: $\mu > 208$    (claim)

C. V. $= 2.896$    $\overline{X} = 209.74$    $s = 1.67$
$$t = \frac{\overline{X} - \mu}{\frac{s}{\sqrt{n}}} = \frac{209.74 - 208}{\frac{1.67}{\sqrt{9}}} = 3.126$$

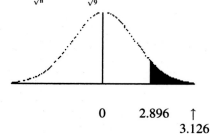

$0 \qquad 2.896 \quad \uparrow$
$\qquad\qquad\qquad 3.126$

7. continued

Reject the null hypothesis. There is enough evidence to support the claim that weight is greater than 208 grams.

9.
$H_0$: $p = 0.602$
$H_1$: $p > 0.602$ (claim)

C. V. $= 1.65$
$\hat{p} = 0.65$    $p = 0.602$    $q = 0.398$

$z = \dfrac{\hat{p} - p}{\sqrt{\frac{pq}{n}}} = \dfrac{0.65 - 0.602}{\sqrt{\frac{(0.602)(0.398)}{400}}} = 1.96$

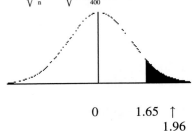

0        1.65  ↑
              1.96

Reject the null hypothesis. There is enough evidence to support the claim that the percentage of drug offenders is higher than 60.2%.

11.
$H_0$: $p = 0.65$ (claim)
$H_1$: $p \neq 0.65$

$\hat{p} = \dfrac{57}{80} = 0.7125$    $p = 0.65$    $q = 0.35$

$z = \dfrac{\hat{p} - p}{\sqrt{\frac{pq}{n}}} = \dfrac{0.7125 - 0.65}{\sqrt{\frac{(0.65)(0.35)}{80}}} = 1.17$

Area $= 0.8790$
P-value $= 2(1 - 0.8790) = 0.242\ (0.2412)$
Since P-value $> 0.05$, do not reject the null hypothesis. There is not enough evidence to reject the claim that 65% of the teenagers own their own radios.

13.
$H_0$: $\mu = 10$
$H_1$: $\mu < 10$ (claim)

$z = \dfrac{\overline{X} - \mu}{\frac{\sigma}{\sqrt{n}}} = \dfrac{9.25 - 10}{\frac{2}{\sqrt{35}}} = -2.22$

Area $= 0.0132$
P-value $= 0.0132$
Since $0.0132 < 0.05$, reject the null hypothesis. The conclusion is that there is

13. continued
enough evidence to support the claim that the average time is less than 10 minutes.

15.
$H_0$: $\sigma = 4.3$ (claim)
$H_1$: $\sigma < 4.3$

d. f. $= 19$
$\chi^2 = \dfrac{(n-1)s^2}{\sigma^2} = \dfrac{(20-1)(2.6)^2}{(4.3)^2} = 6.95$

$0.005 <$ P-value $< 0.01$ (0.006)
Since P-value $< 0.05$, reject the null hypothesis. There is enough evidence to reject the claim that the standard deviation is greater than or equal to 4.3 miles per gallon.

17.
$H_0$: $\sigma^2 = 40$
$H_1$: $\sigma^2 \neq 40$ (claim)

$s^2 = 43$
C. V. $= 19.023$ and $2.700$    d. f. $= 9$

$\chi^2 = \dfrac{(n-1)s^2}{\sigma^2} = \dfrac{(10-1)(43)}{40} = 9.68$

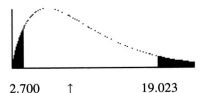

2.700        ↑              19.023
           9.68

Do not reject the null hypothesis. There is not enough evidence to support the claim that the variance is 40.

19.
$H_0$: $\mu = 4$
$H_1$: $\mu \neq 4$ (claim)

C. V. $= \pm 2.58$
$z = \dfrac{\overline{X} - \mu}{\frac{s}{\sqrt{n}}} = \dfrac{4.2 - 4}{\frac{0.6}{\sqrt{20}}} = 1.49$

The 99% confidence interval of the mean is:

$$\overline{X} - z_{\frac{\alpha}{2}} \frac{\sigma}{\sqrt{n}} < \mu < \overline{X} + z_{\frac{\alpha}{2}} \frac{\sigma}{\sqrt{n}}$$

$4.2 - 2.58 \cdot \dfrac{0.6}{\sqrt{20}} < \mu < 4.2 + 2.58 \cdot \dfrac{0.6}{\sqrt{20}}$
$3.85 < \mu < 4.55$

19. continued
The decision is do not reject the null hypothesis since $1.49 < 2.58$ and the confidence interval does contain the hypothesized mean of 4. There is not enough evidence to support the claim that the growth has changed.

CHAPTER 8 QUIZ

1. True
2. True
3. False, the critical value separates the critical region from the noncritical region.
4. True
5. False, it can be one-tailed or two-tailed.
6. b.
7. d.
8. c.
9. b.
10. type I
11. $\beta$
12. statistical hypothesis
13. right
14. $n - 1$

15. $H_0$: $\mu = 28.6$ (claim)
$H_1$: $\mu \neq 28.6$
C. V. $= \pm 1.96$
$z = 2.14$
Reject the null hypothesis. There is enough evidence to reject the claim that the average age is 28.6.

16. $H_0$: $\mu = \$6,500$ (claim)
$H_1$: $\mu \neq \$6,500$
C. V. $= \pm 1.96$
$z = 5.27$
Reject the null hypothesis. There is enough evidence to reject the agent's claim.

17. $H_0$: $\mu = 8$
$H_1$: $\mu > 8$ (claim)
C. V. $= 1.65$
$z = 6.00$
Reject the null hypothesis. There is enough evidence to support the claim that the average number of sticks is greater than 8.

18. $H_0$: $\mu = 500$ (claim)
$H_1$: $\mu \neq 500$
C. V. $= \pm 3.707$
$t = -0.571$

18. continued
Do not reject the null hypothesis. There is not enough evidence to reject the claim that the average is 500.

19. $H_0$: $\mu = 67$
$H_1$: $\mu < 67$ (claim)
$t = -3.1568$
P-value $< 0.005$ (0.003)
Since P-value $< 0.05$, reject the null hypothesis. There is enough evidence to support the claim that the average height is less than 67 inches.

20. $H_0$: $\mu = 12.4$
$H_1$: $\mu < 12.4$ (claim)
C. V. $= -1.345$
$t = -2.324$
Reject the null hypothesis. There is enough evidence to support the claim that the average is less than what the company claimed.

21. $H_0$: $\mu = 63.5$
$H_1$: $\mu > 63.5$ (claim)
$t = 0.47075$
P-value $> 0.25$ (0.322)
Since P-value $> 0.05$, do not reject the null hypothesis. There is not enough evidence to support the claim that the average is greater than 63.5.

22. $H_0$: $\mu = 26$ (claim)
$H_1$: $\mu \neq 26$
C. V. $= \pm 2.492$
$t = -1.5$
Do not reject the null hypothesis. There is not enough evidence to reject the claim that the average age is 26.

23. $H_0$: $p = 0.39$ (claim)
$H_1$: $p \neq 0.39$
C. V. $= \pm 1.96$
$z = -0.62$
Do not reject the null hypothesis. There is not enough evidence to reject the claim that 39% took supplements. The study supports the results of the previous study.

24. $H_0$: $p = 0.55$ (claim)
$H_1$: $p < 0.55$
C. V. $= -1.28$
$z = -0.899$

24. continued
Do not reject the null hypothesis. There is not enough evidence to reject the dietitian's claim.

25. $H_0$: $p = 0.35$     (claim)
$H_1$: $p \neq 0.35$
C. V. $= \pm 2.33$
$z = 0.666$
Do not reject the null hypothesis. There is not enough evidence to reject the claim that the proportion is 35%.

26. $H_0$: $p = 0.75$     (claim)
$H_1$: $p \neq 0.75$
C. V. $= \pm 2.58$
$z = 2.6833$
Reject the null hypothesis. there is enough evidence to reject the claim.

27. The area corresponding to $z = 2.14$ is 0.9838.
P-value $= 2(1 - 0.9838) = 0.0324$

28. The area corresponding to $z = 5.27$ is greater than 0.9999.
Thus, P-value $\leq 2(1 - 0.9999) \leq 0.0002$.
(TI: P-value $= 0.0001$)

29. $H_0$: $\sigma = 6$
$H_1$: $\sigma > 6$     (claim)
C. V. $= 36.415$
$\chi^2 = 54$
Reject the null hypothesis. There is enough evidence to support the claim that the standard deviation is more than 6 pages.

30. $H_0$: $\sigma = 8$     (claim)
$H_1$: $\sigma \neq 8$
C. V. $= 27.991, 79.490$
$\chi^2 = 33.2$
Do not reject the null hypothesis. There is not enough evidence to reject the claim that $\sigma = 8$.

31. $H_0$: $\sigma = 2.3$
$H_1$: $\sigma < 2.3$     (claim)
C. V. $= 10.117$
$\chi^2 = 13$
Do not reject the null hypothesis. There is not enough evidence to support the claim that the standard deviation is less than 2.3.

32. $H_0$: $\sigma = 9$     (claim)
$H_1$: $\sigma \neq 9$

32. continued
$\chi^2 = 13.4$
P-value $> 0.20$     (0.290)
Since P-value $> 0.05$, do not reject the null hypothesis. There is not enough evidence to reject the claim that $\sigma = 9$.

33. $28.9 < \mu < 31.2$; no

34. $\$6562.81 < \mu < \$6,637.19$; no

Note: Graphs are not to scale and are intended to convey a general idea.
Answers may vary due to rounding, TI-83's, or computer programs.

**EXERCISE SET 9-1**

**1.**
Testing a single mean involves comparing a sample mean to a specific value such as $\mu = 100$; whereas testing the difference between means means comparing the means of two samples such as $\mu_1 = \mu_2$.

**3.**
The populations must be independent of each other and they must be normally distributed. $s_1$ and $s_2$ can be used in place of $\sigma_1$ and $\sigma_2$ when $\sigma_1$ and $\sigma_2$ are unknown and both samples are each greater than or equal to 30.

**5.**
$H_0: \mu_1 = \mu_2$  (claim)
$H_1: \mu_1 \neq \mu_2$

C. V. $= \pm 2.58$

$\overline{X}_1 = 662.6111 \qquad \overline{X}_2 = 758.875$

$$z = \frac{(\overline{X}_1 - \overline{X}_2) - (\mu_1 - \mu_2)}{\sqrt{\frac{\sigma_1^2}{n_1} + \frac{\sigma_2^2}{n_2}}} = \frac{(662.6111 - 758.875) - 0}{\sqrt{\frac{450^2}{36} + \frac{474^2}{36}}} =$$

$z = -0.88$
(TI83 answer is $z = -0.856$)

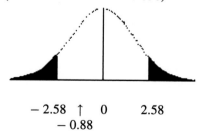

$$-2.58 \uparrow \quad 0 \qquad 2.58$$
$$-0.88$$

Do not reject the null hypothesis. There is not enough evidence to reject the claim that the average lengths of the rivers is the same.

**7.**
$H_0: \mu_1 = \mu_2$
$H_1: \mu_1 \neq \mu_2$  (claim)

C. V. $= \pm 1.96$

**7. continued**

$$z = \frac{(\overline{X}_1 - \overline{X}_2) - (\mu_1 - \mu_2)}{\sqrt{\frac{\sigma_1^2}{n_1} + \frac{\sigma_2^2}{n_2}}} = \frac{(28.5 - 35.2) - 0}{\sqrt{\frac{7.2^2}{40} + \frac{9.1^2}{40}}} = -3.65$$

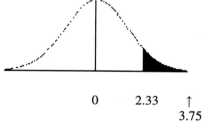

$$\uparrow \quad -1.96 \qquad 0 \qquad 1.96$$
$$-3.65$$

Reject the null hypothesis. There is enough evidence to support the claim commuting times are different in the winter.

**9.**
$H_0: \mu_1 = \mu_2$
$H_1: \mu_1 > \mu_2$  (claim)

C. V. $= 2.33$

$$z = \frac{(\overline{X}_1 - \overline{X}_2) - (\mu_1 - \mu_2)}{\sqrt{\frac{\sigma_1^2}{n_1} + \frac{\sigma_2^2}{n_2}}} = \frac{(5.5 - 4.2) - 0}{\sqrt{\frac{1.2^2}{32} + \frac{1.5^2}{30}}} = 3.75$$

$$0 \qquad 2.33 \quad \uparrow$$
$$3.75$$

Reject the null hypothesis. There is enough evidence to support the claim that the average stay is longer for men than for women.

**11.**
$H_0: \mu_1 = \mu_2$
$H_1: \mu_1 < \mu_2$  (claim)

C. V. $= -1.65$

$$z = \frac{(\overline{X}_1 - \overline{X}_2) - (\mu_1 - \mu_2)}{\sqrt{\frac{\sigma_1^2}{n_1} + \frac{\sigma_2^2}{n_2}}} = \frac{(3.16 - 3.28) - 0}{\sqrt{\frac{0.52^2}{103} + \frac{0.46^2}{225}}} = -2.01$$

**11. continued**

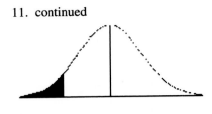

$$\uparrow \quad -1.65 \quad 0$$
$$-2.01$$

Reject the null hypothesis. There is enough evidence to support the claim that those who stayed have a higher GPA than those who left their profession.

**13.**
$H_0$: $\mu_1 = \mu_2$
$H_1$: $\mu_1 > \mu_2$    (claim)

C. V. = 2.33
$\overline{X}_1 = \$9224$        $\overline{X}_2 = \$8497.5$
$s_1 = 3829.826$      $s_2 = 2745.293$

$$z = \frac{(\overline{X}_1 - \overline{X}_2) - (\mu_1 - \mu_2)}{\sqrt{\frac{\sigma_1^2}{n_1} + \frac{\sigma s_2^2}{n_2}}}$$

$$z = \frac{(9224 - 8497.5) - 0}{\sqrt{\frac{3830^2}{50} + \frac{2745^2}{50}}} = 1.09$$

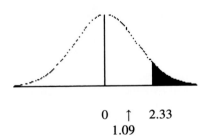

$$0 \quad \uparrow \quad 2.33$$
$$1.09$$

Do not reject the null hypothesis. There is not enough evidence to support the claim that colleges spent more money on men's sports than women's.

**15.**
$H_0$: $\mu_1 = \mu_2$
$H_1$: $\mu_1 \neq \mu_2$    (claim)

$$z = \frac{(\overline{X}_1 - \overline{X}_2) - (\mu_1 - \mu_2)}{\sqrt{\frac{\sigma_1^2}{n_1} + \frac{\sigma_2^2}{n_2}}} = \frac{(3.05 - 2.96) - 0}{\sqrt{\frac{0.75^2}{103} + \frac{0.75^2}{225}}}$$

$z = 1.01$
Area = 0.8438
P-value = $2(1 - 0.8438) = 0.3124$

**15. continued**
Since P-value > 0.05, do not reject the null hypothesis. There is not enough evidence to support the claim that there is a difference in scores. (TI: P-value = 0.3131)

**17.**
$\overline{D} = 83.6 - 79.2 = 4.4$

$$(\overline{X}_1 - \overline{X}_2) - z_{\frac{\alpha}{2}} \sqrt{\frac{\sigma_1^2}{n_1} + \frac{\sigma_2^2}{n_2}} < \mu_1 - \mu_2 <$$
$$(\overline{X}_1 - \overline{X}_2) + z_{\frac{\alpha}{2}} \sqrt{\frac{\sigma_1^2}{n_1} + \frac{\sigma_2^2}{n_2}}$$

$$4.4 - (1.65)\sqrt{\frac{4.3^2}{36} + \frac{3.8^2}{36}} < \mu_1 - \mu_2 <$$
$$4.4 + (1.65)\sqrt{\frac{4.3^2}{36} + \frac{3.8^2}{36}}$$

$2.8 < \mu_1 - \mu_2 < 6.0$
(TI: $2.83 < \mu_1 - \mu_2 < 5.97$)

**19.**
$\overline{D} = 315 - 280 = 35$

$$(\overline{X}_1 - \overline{X}_2) - z_{\frac{\alpha}{2}} \sqrt{\frac{\sigma_1^2}{n_1} + \frac{\sigma_2^2}{n_2}} < \mu_1 - \mu_2 <$$
$$(\overline{X}_1 - \overline{X}_2) + z_{\frac{\alpha}{2}} \sqrt{\frac{\sigma_1^2}{n_1} + \frac{\sigma_2^2}{n_2}}$$

$$35 - (1.96)\sqrt{\frac{56.2^2}{40} + \frac{52.1^2}{35}} < \mu_1 - \mu_2 <$$
$$35 + (1.96)\sqrt{\frac{56.2^2}{40} + \frac{52.1^2}{35}}$$

$10.48 < \mu_1 - \mu_2 < 59.52$

The interval gives evidence to reject the claim that there is no difference in the means because 0 is not contained in the interval.

**21.**
$H_0$: $\mu_1 - \mu_2 = 8$    (claim)
$H_1$: $\mu_1 - \mu_2 > 8$

C. V. = 1.65
$$z = \frac{(\overline{X}_1 - \overline{X}_2) - K}{\sqrt{\frac{\sigma_1^2}{n_1} + \frac{\sigma_2^2}{n_2}}} = \frac{(110 - 104) - 8}{\sqrt{\frac{15^2}{60} + \frac{15^2}{60}}} = -0.73$$

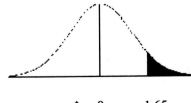

$$\uparrow \quad 0 \quad\quad 1.65$$
$$-0.73$$

21. continued
Do not reject the null hypothesis. There is
not enough evidence to reject the claim that
private school students have exam scores
that are at most 8 points higher than public
school students.

## EXERCISE SET 9-2

1.
$H_0$: $\mu_1 = \mu_2$
$H_1$: $\mu_1 \neq \mu_2$   (claim)

C. V. = $\pm 2.262$   d. f. = 9

$t = \frac{(\overline{X}_1 - \overline{X}_2) - (\mu_1 - \mu_2)}{\sqrt{\frac{s_1^2}{n_1} + \frac{s_2^2}{n_2}}}$

$t = \frac{(83,256 - 88,354) - 0}{\sqrt{\frac{3256^2}{10} + \frac{2341^2}{10}}} = -4.02$

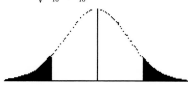

↑ $-2.262$      0      2.262
$-4.02$

Reject the null hypothesis. There is enough
evidence to support the claim that there is a
significant difference in the values of the
homes based upon the appraisers' values.

95% Confidence Interval:
$-5098 - 2.262\sqrt{\frac{3256^2}{10} + \frac{2341^2}{10}} < \mu_1 - \mu_2$
$< -5098 + 2.262\sqrt{\frac{3256^2}{10} + \frac{2341^2}{10}}$
$-\$7967 < \mu_1 - \mu_2 < -\$2230$
(TI:  $-\$7782 < \mu_1 - \mu_2 < -\$2414$)

3.
$H_0$: $\mu_1 = \mu_2$   (claim)
$H_1$: $\mu_1 \neq \mu_2$

C. V. = $\pm 2.145$
d. f. = 14

$t = \frac{(\overline{X}_1 - \overline{X}_2) - (\mu_1 - \mu_2)}{\sqrt{\frac{s_1^2}{n_1} + \frac{s_2^2}{n_2}}}$

$t = \frac{(501,580 - 513,360) - 0}{\sqrt{\frac{20,000^2}{15} + \frac{18,000^2}{15}}} = -1.70$

3. continued

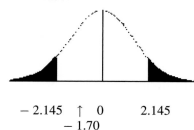

$-2.145$ ↑ 0      2.145
$-1.70$

Do not reject the null hypothesis. There is
enough evidence to support the claim that
there is no difference between the salaries.

5.
$H_0$: $\mu_1 = \mu_2$
$H_1$: $\mu_1 > \mu_2$   (claim)

C. V. = 1.729      d. f. = 19

$t = \frac{(\overline{X}_1 - \overline{X}_2) - (\mu_1 - \mu_2)}{\sqrt{\frac{s_1^2}{n_1} + \frac{s_2^2}{n_2}}}$

$t = \frac{(62.1 - 55.6) - 0}{\sqrt{\frac{5.4^2}{20} + \frac{3.9^2}{20}}} = 4.36$

P-value = 0.00 (0.00005) which is $< \alpha$.
(TI: P-value = 0.00005)
Reject the null hypothesis. There is enough
evidence to support the claim that houses in
Whiting are older.

7.
$H_0$: $\mu_1 = \mu_2$
$H_1$: $\mu_1 \neq \mu_2$   (claim)

C. V. = $\pm 2.821$      d. f. = 9

$t = \frac{(\overline{X}_1 - \overline{X}_2) - (\mu_1 - \mu_2)}{\sqrt{\frac{s_1^2}{n_1} + \frac{s_2^2}{n_2}}}$

$t = \frac{(21 - 27) - 0}{\sqrt{\frac{5.6^2}{10} + \frac{4.3^2}{14}}} = -2.84$

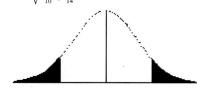

↑ $-2.821$      0      2.821
$-2.84$

**7. continued**

Reject the null hypothesis. There is enough evidence to support the claim that there is a difference in the average times of the two groups.

98% Confidence Interval:

$$-6 - 2.821\sqrt{\frac{5.6^2}{10} + \frac{4.3^2}{14}} < \mu_1 - \mu_2$$

$$< -6 + 2.821\sqrt{\frac{5.6^2}{10} + \frac{4.3^2}{14}}$$

$-11.96 < \mu_1 - \mu_2 < -0.04$
(TI: $-11.5 < \mu_1 - \mu_2 < -0.55$)

**9.**

$H_0$: $\mu_1 = \mu_2$
$H_1$: $\mu_1 \neq \mu_2$   (claim)

$\overline{X}_1 = 82.875$    $s_1 = 6.8959$
$\overline{X}_2 = 91.667$    $s_2 = 3.7417$
C.V. $= \pm 2.365$  df $= 7$

$$t = \frac{(\overline{X}_1 - \overline{X}_2) - (\mu_1 - \mu_2)}{\sqrt{\frac{s_1^2}{n_1} + \frac{s_2^2}{n_2}}}$$

$$t = \frac{(82.875 - 91.667) - 0}{\sqrt{\frac{6.8959^2}{8} + \frac{3.7417^2}{9}}} = -3.21$$

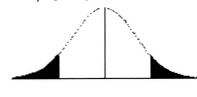

↑   $-2.365$    0    2.365
$-3.21$

Reject the null hypothesis. There is enough evidence to support the claim that fruits and vegetables differ in moisture content.

**11.**

$H_0$: $\mu_1 = \mu_2$
$H_1$: $\mu_1 \neq \mu_2$   (claim)

$\overline{X}_1 = 65.7273$    $s_1 = 9.1224$
$\overline{X}_2 = 60.2222$    $s_2 = 11.3883$

C. V. $= \pm 2.306$     d. f. $= 8$

$$t = \frac{(65.7273 - 60.2222) - 0}{\sqrt{\frac{9.1224^2}{11} + \frac{11.3883^2}{9}}} = 1.17$$

**11. continued**

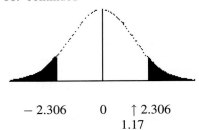

$-2.306$      0    ↑ 2.306
           1.17

Do not reject the null hypothesis. There is not enough evidence to support the claim that there is a difference in the means.

**13.**
Private: $\overline{X} = \$16,147.5$    s $= 4023.7$
Public: $\overline{X} = \$9039.9$      s $= 3325.5$

95% Confidence Interval:

$$7107.6 - 2.571\sqrt{\frac{4023.7^2}{6} + \frac{3325.5^2}{7}}$$

$$< \mu_1 - \mu_2 <$$

$$7107.6 + 2.571\sqrt{\frac{4023.7^2}{6} + \frac{3325.5^2}{7}}$$

$\$1789.70 < \mu_1 - \mu_2 < \$12,425.41$
(TI: $\$2484.6 < \mu_1 - \mu_2 < \$11,731$)

**EXERCISE SET 9-3**

**1.**
a. dependent
b. dependent
c. independent
d. dependent
e. independent

**3.**

| Before | After | D | D$^2$ |
|--------|-------|------|------|
| 9 | 9 | 0 | 0 |
| 12 | 17 | -5 | 25 |
| 6 | 9 | -3 | 9 |
| 15 | 20 | -5 | 25 |
| 3 | 2 | 1 | 1 |
| 18 | 21 | -3 | 9 |
| 10 | 15 | -5 | 25 |
| 13 | 22 | -9 | 81 |
| 7 | 6 | 1 | 1 |
| | | $\sum D = -28$ | $\sum D^2 = 176$ |

$H_0$: $\mu_D = 0$
$H_1$: $\mu_D < 0$   (claim)

C. V. $= -1.397$       d. f. $= 8$

3. continued

$\overline{D} = \frac{\sum D}{n} = -3.11$

$s_D = \sqrt{\frac{n\sum D^2 - (\sum D)^2}{n(n-1)}} = \sqrt{\frac{9(176) - (-28)^2}{9(8)}} = 3.3$

$t = \frac{-3.11 - 0}{\frac{3.33}{\sqrt{9}}} = -2.8$

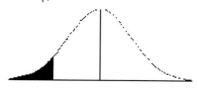

$\uparrow -1.397 \qquad 0$
$-2.8$

Reject the null hypothesis. There is enough evidence to support the claim that the seminar increased the number of hours students studied.

5.

| F - S | S - Th | D | D $^2$ |
|---|---|---|---|
| 4 | 8 | -4 | 16 |
| 7 | 5.5 | 1.5 | 2.25 |
| 10.5 | 7.5 | 3 | 9 |
| 12 | 8 | 4 | 16 |
| 11 | 7 | 4 | 16 |
| 9 | 6 | 3 | 9 |
| 6 | 6 | 0 | 0 |
| 9 | 8 | 1 | 1 |

$\sum D = 12.5 \quad \sum D^2 = 69.25$

$H_0$: $\mu_D = 0$
$H_1$: $\mu_D \neq 0$  (claim)

C. V. $= \pm 2.365$ \qquad d. f. $= 7$

$\overline{D} = \frac{\sum D}{n} = \frac{12.5}{8} = 1.5625$

$s_D = \sqrt{\frac{n\sum D^2 - (\sum D)^2}{n(n-1)}}$

$s_D = \sqrt{\frac{8(69.25) - (12.5)^2}{8(7)}} = 2.665$

$t = \frac{1.5625 - 0}{\frac{2.665}{\sqrt{8}}} = 1.6583$

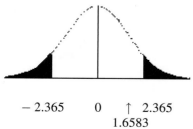

$-2.365 \qquad 0 \qquad \uparrow 2.365$
$1.6583$

5. continued

Do not reject the null hypothesis. There is not enough evidence to support the claim that there is a difference in the mean number of hours slept.

7.

| Before | After | D | D $^2$ |
|---|---|---|---|
| 12 | 9 | 3 | 9 |
| 9 | 6 | 3 | 9 |
| 0 | 1 | -1 | 1 |
| 5 | 3 | 2 | 4 |
| 4 | 2 | 2 | 4 |
| 3 | 3 | 0 | 0 |

$\sum D = 9 \quad \sum D^2 = 27$

$H_0$: $\mu_D = 0$
$H_1$: $\mu_D > 0$  (claim)

C. V. $= 2.571$ \qquad d. f. $= 5$

$\overline{D} = \frac{\sum D}{n} = \frac{9}{6} = 1.5$

$s_D = \sqrt{\frac{n\sum D^2 - (\sum D)^2}{n(n-1)}} = \sqrt{\frac{6(27) - 9^2}{6(5)}} = 1.64$

$t = \frac{1.5 - 0}{\frac{1.64}{\sqrt{6}}} = 2.24$

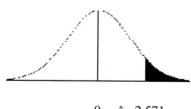

$0 \quad \uparrow 2.571$
$2.24$

Do not reject the null hypothesis. There is not enough evidence to support the claim that the errors have been reduced.

9.

| A | B | D | D $^2$ |
|---|---|---|---|
| 87 | 83 | 4 | 16 |
| 92 | 95 | -3 | 9 |
| 78 | 79 | -1 | 1 |
| 83 | 83 | 0 | 0 |
| 88 | 86 | 2 | 4 |
| 90 | 93 | -3 | 9 |
| 84 | 80 | 4 | 16 |
| 93 | 86 | 7 | 49 |

$\sum D = 10 \quad \sum D^2 = 104$

$H_0$: $\mu_D = 0$
$H_1$: $\mu_D \neq 0$  (claim)

**9. continued**

P-value = 0.361          d. f. = 7

$$\overline{D} = \frac{\sum D}{n} = \frac{10}{8} = 1.25$$

$$s_D = \sqrt{\frac{n\sum D^2 - (\sum D)^2}{n(n-1)}} = \sqrt{\frac{8(104) - 10^2}{8(7)}} = 3.62$$

$$t = \frac{1.25 - 0}{\frac{3.62}{\sqrt{8}}} = 0.978$$

0.20 < P-value < 0.50  Do not reject the null hypothesis since P-value > 0.01.  There is not enough evidence to support the claim that there is a difference in the pulse rates.

Confidence Interval:

$$1.25 - 3.499\left(\frac{3.62}{\sqrt{8}}\right) < \mu_D <$$
$$1.25 + 3.499\left(\frac{3.62}{\sqrt{8}}\right)$$
$$-3.23 < \mu_D < 5.73$$

**11.**

Using the previous problem, $\overline{D} = -1.5625$ whereas the mean of the 1994 values is 95.375 and the mean of the 1999 values is 96.9375; hence,
$$\overline{D} = 95.375 - 96.9375 = -1.5625$$

## EXERCISE SET 9-4

**1A.**

Use $\hat{p} = \frac{X}{n}$ and $\hat{q} = 1 - \hat{p}$

a. $\hat{p} = \frac{34}{48}$          $\hat{q} = \frac{14}{48}$

b. $\hat{p} = \frac{28}{75}$          $\hat{q} = \frac{47}{75}$

c. $\hat{p} = \frac{50}{100}$          $\hat{q} = \frac{50}{100}$

d. $\hat{p} = \frac{6}{24}$          $\hat{q} = \frac{18}{24}$

e. $\hat{p} = \frac{12}{144}$          $\hat{q} = \frac{132}{144}$

**1B.**

a. $x = 0.16(100) = 16$
b. $x = 0.08(50) = 4$
c. $x = 0.06(800) = 48$
d. $x = 0.52(200) = 104$
e. $x = 0.20(150) = 30$

**3.**

$$\hat{p}_1 = \frac{X_1}{n_1} = \frac{98}{300} = 0.3267 \quad \hat{p}_2 = \frac{120}{300} = 0.4$$

**3. continued**

$$\overline{p} = \frac{X_1 + X_2}{n_1 + n_2} = \frac{98 + 120}{300 + 300} = \frac{218}{600} = 0.363$$
$$\overline{q} = 1 - \overline{p} = 1 - 0.363 = 0.637$$

$H_0$: $p_1 = p_2$
$H_1$: $p_1 \neq p_2$     (claim)

C. V. = $\pm 2.58$

$$z = \frac{(\hat{p}_1 - \hat{p}_2) - (p_1 - p_2)}{\sqrt{(\overline{p})(\overline{q})\left(\frac{1}{n_1} + \frac{1}{n_2}\right)}} = \frac{(0.3267 - 0.4) - 0}{\sqrt{(0.363)(0.637)\left(\frac{1}{300} + \frac{1}{300}\right)}}$$

$z = -1.87$
(TI: $z = -1.867$)

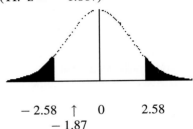

$$-2.58 \quad \uparrow \quad 0 \qquad 2.58$$
$$-1.87$$

Do not reject the null hypothesis.  There is not enough evidence to support the claim that there is a significant difference in the proportions.

**5.**

$$\hat{p}_1 = \frac{X_1}{n_1} = \frac{112}{150} = 0.7467 \quad \hat{p}_2 = \frac{150}{200} = 0.75$$

$$\overline{p} = \frac{X_1 + X_2}{n_1 + n_2} = \frac{112 + 150}{150 + 200} = 0.749$$
$$\overline{q} = 1 - \overline{p} = 1 - 0.749 = 0.251$$

$H_0$: $p_1 = p_2$
$H_1$: $p_1 \neq p_2$     (claim)

C. V. = $\pm 1.96$

$$z = \frac{(\hat{p}_1 - \hat{p}_2) - (p_1 - p_2)}{\sqrt{(\overline{p})(\overline{q})\left(\frac{1}{n_1} + \frac{1}{n_2}\right)}} = \frac{(0.7467 - 0.75) - 0}{\sqrt{(0.749)(0.251)\left(\frac{1}{150} + \frac{1}{200}\right)}}$$

$z = -0.07$

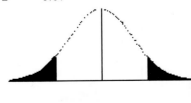

$$-1.96 \quad \uparrow 0 \qquad 1.96$$
$$-0.07$$

**5. continued**

Do not reject the null hypothesis. There is not enough evidence to support the claim that the proportions are different.

**7.**

$\hat{p}_1 = 0.83 \qquad \hat{p}_2 = 0.75$

$X_1 = 0.83(100) = 83$

$X_2 = 0.75(100) = 75$

$\bar{p} = \frac{83 + 75}{100 + 100} = 0.79 \qquad \bar{q} = 1 - 0.79 = 0.21$

$H_0$: $p_1 = p_2$ (claim)

$H_1$: $p_1 \neq p_2$

C. V. $= \pm 1.96 \qquad \alpha = 0.05$

$z = \frac{(\hat{p}_1 - \hat{p}_2) - (p_1 - p_2)}{\sqrt{(\bar{p})(\bar{q})\left(\frac{1}{n_1} + \frac{1}{n_2}\right)}} = \frac{(0.83 - 0.75) - 0}{\sqrt{(0.79)(0.21)\left(\frac{1}{100} + \frac{1}{100}\right)}}$

$z = 1.39$

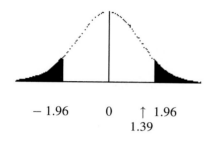

$-1.96 \qquad 0 \qquad \uparrow 1.96$
$\qquad\qquad\qquad 1.39$

Do not reject the null hypothesis. There is not enough evidence to reject the claim that the proportions are equal.

$(\hat{p}_1 - \hat{p}_2) - z_{\frac{\alpha}{2}} \sqrt{\frac{\hat{p}_1\hat{q}_1}{n_1} + \frac{\hat{p}_2\hat{q}_2}{n_2}} < p_1 - p_2 <$

$\qquad (\hat{p}_1 - \hat{p}_2) + z_{\frac{\alpha}{2}} \sqrt{\frac{\hat{p}_1\hat{q}_1}{n_1} + \frac{\hat{p}_2\hat{q}_2}{n_2}}$

$0.08 - 1.96 \sqrt{\frac{0.83(0.17)}{100} + \frac{0.75(0.25)}{100}} < p_1 - p_2$

$\qquad < 0.08 + 1.96 \sqrt{\frac{0.83(0.17)}{100} + \frac{0.75(0.25)}{100}}$

$-0.032 < p_1 - p_2 < 0.192$

**9.**

$\hat{p}_1 = 0.55 \qquad \hat{p}_2 = 0.45$

$X_1 = 0.55(80) = 44 \quad X_2 = 0.45(90) = 40.5$

$\bar{p} = \frac{X_1 + X_2}{n_1 + n_2} = \frac{44 + 40.5}{80 + 90} = 0.497$

**9. continued**

$\bar{q} = 1 - \bar{p} = 1 - 0.497 = 0.503$

$H_0$: $p_1 = p_2$

$H_1$: $p_1 \neq p_2$ (claim)

C. V. $= \pm 2.58 \qquad \alpha = 0.01$

$z = \frac{(\hat{p}_1 - \hat{p}_2) - (p_1 - p_2)}{\sqrt{(\bar{p})(\bar{q})\left(\frac{1}{n_1} + \frac{1}{n_2}\right)}} = \frac{(0.55 - 0.45) - 0}{\sqrt{(0.497)(0.503)\left(\frac{1}{80} + \frac{1}{90}\right)}}$

$z = 1.302$

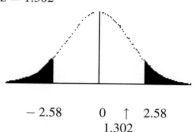

$-2.58 \qquad 0 \quad \uparrow \quad 2.58$
$\qquad\qquad\qquad 1.302$

Do not reject the null hypothesis. There is not enough evidence to support the claim that the proportions are different.

$(\hat{p}_1 - \hat{p}_2) - z_{\frac{\alpha}{2}} \sqrt{\frac{\hat{p}_1\hat{q}_1}{n_1} + \frac{\hat{p}_2\hat{q}_2}{n_2}} < p_1 - p_2 <$

$\qquad (\hat{p}_1 - \hat{p}_2) + z_{\frac{\alpha}{2}} \sqrt{\frac{\hat{p}_1\hat{q}_1}{n_1} + \frac{\hat{p}_2\hat{q}_2}{n_2}}$

$0.1 - 2.58 \sqrt{\frac{0.55(0.45)}{80} + \frac{0.45(0.55)}{90}} < p_1 - p_2$

$\qquad < 0.1 + 2.58 \sqrt{\frac{0.55(0.45)}{80} + \frac{0.45(0.55)}{90}}$

$-0.097 < p_1 - p_2 < 0.297$

**11.**

$\hat{p}_1 = \frac{26}{75} = 0.3467 \qquad \hat{p}_2 = \frac{26}{60} = 0.4333$

$\bar{p} = \frac{X_1 + X_2}{n_1 + n_2} = \frac{26 + 26}{75 + 60} = 0.3852$

$\bar{q} = 1 - \bar{p} = 1 - 0.3852 = 0.6148$

$H_0$: $p_1 = p_2$

$H_1$: $p_1 \neq p_2$ (claim)

C. V. $= \pm 1.96 \qquad \alpha = 0.05$

$z = \frac{(\hat{p}_1 - \hat{p}_2) - (p_1 - p_2)}{\sqrt{(\bar{p})(\bar{q})\left(\frac{1}{n_1} + \frac{1}{n_2}\right)}} = \frac{(0.3467 - 0.4333) - 0}{\sqrt{(0.3852)(0.6148)\left(\frac{1}{75} + \frac{1}{60}\right)}}$

$z = -1.027$

11. continued

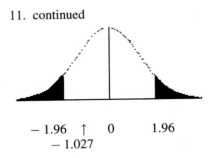

$$-1.96 \quad \uparrow \quad 0 \quad 1.96$$
$$-1.027$$

Do not reject the null hypothesis. There is not enough evidence to support the claim that the proportion of dog owners has changed.

$$(\hat{p}_1 - \hat{p}_2) - z_{\frac{\alpha}{2}} \sqrt{\frac{\hat{p}_1 \hat{q}_1}{n_1} + \frac{\hat{p}_2 \hat{q}_2}{n_2}} < p_1 - p_2 <$$

$$(\hat{p}_1 - \hat{p}_2) + z_{\frac{\alpha}{2}} \sqrt{\frac{\hat{p}_1 \hat{q}_1}{n_1} + \frac{\hat{p}_2 \hat{q}_2}{n_2}}$$

$$-0.0866 - 1.96 \sqrt{\frac{0.3467(0.6533)}{75} + \frac{0.4333(0.5667)}{60}}$$
$$< p_1 - p_2 <$$

$$-0.0866 + 1.96 \sqrt{\frac{0.3467(0.6533)}{75} + \frac{0.4333(0.5667)}{60}}$$

$$-0.252 < p_1 - p_2 < 0.079$$

Yes, the confidence interval contains 0. This is another way to conclude that there is no difference in the proportions.

13.
$$\hat{p}_1 = \frac{X_1}{n_1} = \frac{50}{200} = 0.25$$

$$\hat{p}_2 = \frac{X_2}{n_2} = \frac{93}{300} = 0.31$$

$$\bar{p} = \frac{X_1 + X_2}{n_1 + n_2} = \frac{50 + 93}{200 + 300} = 0.286$$

$$\bar{q} = 1 - \bar{p} = 1 - 0.286 = 0.714$$

$H_0$: $p_1 = p_2$
$H_1$: $p_1 \neq p_2$  (claim)

C.V. $= \pm 2.58$  $\alpha = 0.01$

$$z = \frac{(\hat{p}_1 - \hat{p}_2) - (p_1 - p_2)}{\sqrt{(\bar{p})(\bar{q})(\frac{1}{n_1} + \frac{1}{n_2})}} = \frac{(0.25 - 0.31) - 0}{\sqrt{(0.286)(0.714)(\frac{1}{200} + \frac{1}{300})}}$$

$$z = -1.45$$

13. continued

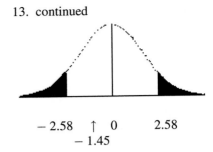

$$-2.58 \quad \uparrow \quad 0 \quad 2.58$$
$$-1.45$$

Do not reject the null hypothesis. There is not enough evidence to support the claim that the proportions are different.

$$(\hat{p}_1 - \hat{p}_2) - z_{\frac{\alpha}{2}} \sqrt{\frac{\hat{p}_1 \hat{q}_1}{n_1} + \frac{\hat{p}_2 \hat{q}_2}{n_2}} < p_1 - p_2 <$$

$$(\hat{p}_1 - \hat{p}_2) + z_{\frac{\alpha}{2}} \sqrt{\frac{\hat{p}_1 \hat{q}_1}{n_1} + \frac{\hat{p}_2 \hat{q}_2}{n_2}}$$

$$-0.06 - 2.58 \sqrt{\frac{0.25(0.75)}{200} + \frac{0.31(0.69)}{300}} <$$

$$p_1 - p_2 < -0.06 + 2.58 \sqrt{\frac{0.25(0.75)}{200} + \frac{0.31(0.69)}{300}}$$

$$-0.165 < p_1 - p_2 < 0.045$$

15.
$\alpha = 0.01$
$\hat{p}_1 = 0.8$  $\qquad \hat{q}_1 = 0.2$
$\hat{p}_2 = 0.6$  $\qquad \hat{q}_2 = 0.4$

$$\hat{p}_1 - \hat{p}_2 = 0.8 - 0.6 = 0.2$$

$$(\hat{p}_1 - \hat{p}_2) - z_{\frac{\alpha}{2}} \sqrt{\frac{\hat{p}_1 \hat{q}_1}{n_1} + \frac{\hat{p}_2 \hat{q}_2}{n_2}} < p_1 - p_2 <$$

$$(\hat{p}_1 - \hat{p}_2) + z_{\frac{\alpha}{2}} \sqrt{\frac{\hat{p}_1 \hat{q}_1}{n_1} + \frac{\hat{p}_2 \hat{q}_2}{n_2}}$$

$$0.2 - 2.58 \sqrt{\frac{(0.8)(0.2)}{150} + \frac{(0.6)(0.4)}{200}} < p_1 - p_2 <$$

$$0.2 + 2.58 \sqrt{\frac{(0.8)(0.2)}{150} + \frac{(0.6)(0.4)}{200}}$$

$$0.077 < p_1 - p_2 < 0.323$$

17.
$$\hat{p}_1 = \frac{X_1}{n_1} = \frac{99}{300} = 0.33 \quad \hat{p}_2 = \frac{81}{300} = 0.27$$

$$\bar{p} = \frac{X_1 + X_2}{n_1 + n_2} = \frac{99 + 81}{300 + 300} = 0.3$$

$$\bar{q} = 1 - \bar{p} = 1 - 0.3 = 0.7$$

17. continued

$H_0$: $p_1 = p_2$

$H_1$: $p_1 > p_2$    (claim)

C. V. = 1.645

$$z = \frac{(\hat{p}_1 - \hat{p}_2) - (p_1 - p_2)}{\sqrt{(\bar{p})(\bar{q})\left(\frac{1}{n_1} + \frac{1}{n_2}\right)}} = \frac{(0.33 - 0.27) - 0}{\sqrt{(0.3)(0.7)\left(\frac{1}{300} + \frac{1}{300}\right)}}$$

$z = 1.60$

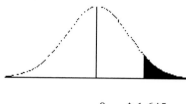

0    ↑ 1.645
     1.60

Do not reject the null hypothesis. There is not enough evidence to support the claim that the proportion of never-married men is greater than the proportion of never-married women.

19.

$\hat{p}_1 = 0.2875$     $\hat{q}_1 = 0.7125$

$\hat{p}_2 = 0.2857$     $\hat{q}_2 = 0.7143$

$\hat{p}_1 - \hat{p}_2 = 0.0018$

$$(\hat{p}_1 - \hat{p}_2) - z_{\frac{\alpha}{2}}\sqrt{\frac{\hat{p}_1\hat{q}_1}{n_1} + \frac{\hat{p}_2\hat{q}_2}{n_2}} < p_1 - p_2 <$$

$$(\hat{p}_1 - \hat{p}_2) + z_{\frac{\alpha}{2}}\sqrt{\frac{\hat{p}_1\hat{q}_1}{n_1} + \frac{\hat{p}_2\hat{q}_2}{n_2}}$$

$$0.0018 - 1.96\sqrt{\frac{(0.2875)(0.7125)}{400} + \frac{(0.2857)(0.7143)}{350}} < p_1 - p_2$$

$$< 0.0018 + 1.96\sqrt{\frac{(0.2875)(0.7125)}{400} + \frac{(0.2857)(0.7143)}{350}}$$

$$-0.0631 < p_1 - p_2 < 0.0667$$

The interval does agree with the *Almanac* statistics stating a difference of $-0.042$ since $-0.042$ is contained in the interval.

## EXERCISE SET 9-5

1.

It should be the larger of the two variances.

3.

One d.f. is used for the variance associated with the numerator and one is used for the variance associated with the denominator.

5.

a. d. f. N = 15, d. f. D = 22; C. V. = 3.36

b. d. f. N = 24, d. f. D = 13; C. V. = 3.59

c. d. f. N = 45, d. f. D = 29; C. V. = 2.03

d. d. f. N = 20, d. f. D = 16; C. V. = 2.28

e. d. f. N = 10, d. f. D = 10; C. V. = 2.98

6.

Note: Specific P-values are in parentheses.

a. $0.025 <$ P-value $< 0.05$   (0.033)

b. $0.05 <$ P-value $< 0.10$   (0.072)

c. P-value $= 0.05$

d. $0.005 <$ P-value $< 0.01$ (0.006)

e. P-value $= 0.05$

f. $P > 0.10$             (0.112)

g. $0.05 <$ P-value $< 0.10$   (0.068)

h. $0.01 <$ P-value $< 0.02$   (0.015)

7.

$H_0$: $\sigma_1^2 = \sigma_2^2$

$H_1$: $\sigma_1^2 \neq \sigma_2^2$    (claim)

C. V. = 2.53      $\alpha = \frac{0.10}{2}$

d. f. N = 14     d. f. D = 14

$F = \frac{s_1^2}{s_2^2} = \frac{13.12^2}{6.17^2} = 4.52$

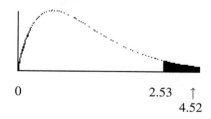

0              2.53   ↑
                    4.52

Reject the null hypothesis. There is enough evidence to support the claim that there is a difference in the variances of the best seller lists for fiction and non-fiction.

9.

$H_0$: $\sigma_1^2 = \sigma_2^2$

$H_1$: $\sigma_1^2 \neq \sigma_2^2$    (claim)

$s_1 = 25.97$      $s_2 = 72.74$

C. V. = 2.86     $\alpha = \frac{0.05}{2}$

d. f. N = 15     d. f. D = 15

$F = \frac{s_1^2}{s_2^2} = \frac{72.74^2}{25.97^2} = 7.85$

9. continued

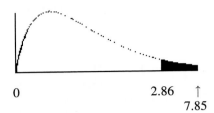

0            2.86   ↑
                  7.85

Reject the null hypothesis. There is enough evidence to support the claim that the variances of the values of tax exempt properties are different. Since both data sets vary greatly from normality, the results are suspect.

11.
$H_0$: $\sigma_1^2 = \sigma_2^2$
$H_1$: $\sigma_1^2 \neq \sigma_2^2$     (claim)

$s_1 = 33.99$      $s_2 = 33.99$
C. V. $= 4.99$     $\alpha = \frac{0.05}{2}$
d. f. N $= 7$      d. f. D $= 7$
$F = \frac{s_1^2}{s_2^2} = \frac{(33.99)^2}{(33.99)^2} = 1$

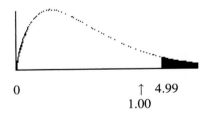

0                ↑   4.99
               1.00

Do not reject the null hypothesis. There is not enough evidence to support the claim that the variance in the number of calories differs between the two brands.

13.
$H_0$: $\sigma_1^2 = \sigma_2^2$
$H_1$: $\sigma_1^2 > \sigma_2^2$     (claim)

$s_1 = 111.211$     $s_2 = 35.523$
$n_1 = 7$         $n_2 = 6$

d. f. N $= 6$      d. f. D $= 5$
C. V. $= 4.950$ at $\alpha = 0.05$
C. V. $= 10.67$ at $\alpha = 0.01$
$F = \frac{s_1^2}{s_2^2} = \frac{(111.211)^2}{(35.523)^2} = 9.801$

13. continued

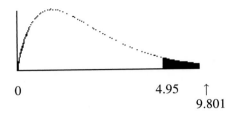

0             4.95    ↑
                 9.801

Reject the null hypothesis at $\alpha = 0.05$. There is enough evidence to support the claim that the variance in area is greater for Eastern cities.

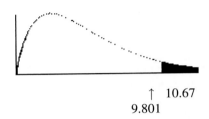

                    ↑   10.67
                9.801

Do not reject the null hypothesis at $\alpha = 0.01$. There is not enough evidence to support the claim that the variance in area is greater for Eastern cities.

15.
$H_0$: $\sigma_1^2 = \sigma_2^2$
$H_1$: $\sigma_1^2 \neq \sigma_2^2$     (claim)

Research: $s_1 = 5501.118$
Primary Care: $s_2 = 5238.809$

C. V. $= 4.03$      $\alpha = \frac{0.05}{2}$
d. f. N $= 9$      d. f. D $= 9$
$F = \frac{s_1^2}{s_2^2} = \frac{(5501.118)^2}{(5238.809)^2} = 1.10$

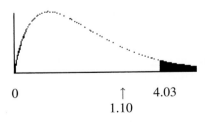

0               ↑     4.03
             1.10

Do not reject the null hypothesis. There is not enough evidence to support the claim that there is a difference between the variances in tuition costs.

# Chapter 9 - Testing the Difference Between
## Two Means, Two Proportions, and Two Variances

**17.**

$H_0$: $\sigma_1^2 = \sigma_2^2$ (claim)
$H_1$: $\sigma_1^2 \neq \sigma_2^2$

$s_1 = 130.496$ $s_2 = 73.215$
C. V. $= 3.87$ $\alpha = \frac{0.10}{2}$
d. f. N $= 6$ d. f. D $= 7$
$F = \frac{s_1^2}{s_2^2} = \frac{(130.496)^2}{(73.215)^2} = 3.18$

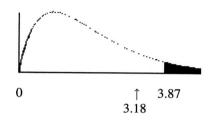

| 0 | ↑ 3.87 |
|---|---|
| | 3.18 |

Do not reject the null hypothesis. There is not enough evidence to reject the claim that the variances of the heights are equal.

**19.**

| Men | Women |
|---|---|
| $s_1^2 = 2.363$ | $s_2^2 = 0.444$ |
| $n_1 = 15$ | $n_2 = 15$ |

$H_0$: $\sigma_1^2 = \sigma_2^2$ (claim)
$H_1$: $\sigma_1^2 \neq \sigma_2$

$\alpha = 0.05$ P-value $= 0.004$
d. f. N $= 14$ d. f. D $= 14$
$F = \frac{s_1^2}{s_2^2} = \frac{2.363}{0.444} = 5.32$

Since P-value $< 0.01$, reject the null hypothesis. There is enough evidence to reject the claim that the variances in weights are equal.

## REVIEW EXERCISES - CHAPTER 9

**1.**

$H_0$: $\mu_1 = \mu_2$
$H_1$: $\mu_1 > \mu_2$ (claim)

CV $= 2.33$ $\alpha = 0.01$
$\overline{X}_1 = 120.1$ $\overline{X}_2 = 117.8$
$s_1 = 16.722$ $s_2 = 16.053$

$z = \frac{(\overline{X}_1 - \overline{X}_2) - (\mu_1 - \mu_2)}{\sqrt{\frac{\sigma_1^2}{n_1} + \frac{\sigma_2^2}{n_2}}} = \frac{(120.1 - 117.8) - 0}{\sqrt{\frac{16.722^2}{36} + \frac{16.053^2}{35}}}$

$z = 0.587$

**1. continued**

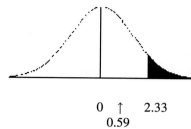

| 0 | ↑ | 2.33 |
|---|---|---|
| | 0.59 | |

Do not reject the null hypothesis. There is not enough evidence to support the claim that single people do more pleasure driving than married people.

**3.**

$H_0$: $\mu_1 = \mu_2$
$H_1$: $\mu_1 > \mu_2$ (claim)

C. V. $= 1.729$ $\alpha = 0.05$

$t = \frac{(\overline{X}_1 - \overline{X}_2) - (\mu_1 - \mu_2)}{\sqrt{\frac{s_1^2}{n_1} + \frac{s_2^2}{n_2}}} = \frac{(16.7 - 12.5) - 0}{\sqrt{\frac{8.41}{26} + \frac{10.24}{20}}}$

$t = 4.595$

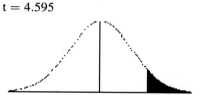

| 0 | 1.729 | ↑ |
|---|---|---|
| | | 4.595 |

Reject the null hypothesis. There is enough evidence to support the claim that single persons spend a greater time each day communicating.

**5.**

$H_0$: $\mu_1 = \mu_2$
$H_1$: $\mu_1 \neq \mu_2$ (claim)

C. V. $= \pm 2.624$ d. f. $= 14$

$t = \frac{(\overline{X}_1 - \overline{X}_2) - (\mu_1 - \mu_2)}{\sqrt{\frac{s_1^2}{n_1} + \frac{s_2^2}{n_2}}} = \frac{(35,270 - 29,512) - 0}{\sqrt{\frac{3256^2}{15} + \frac{1432^2}{15}}}$

$t = 6.54$

**5. continued**

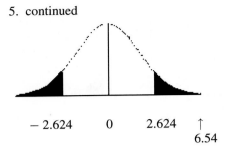

$$-2.624 \qquad 0 \qquad 2.624 \quad \uparrow$$
$$6.54$$

Reject the null hypothesis. There is enough evidence to support the claim that there is a difference in the teachers' salaries.

98% Confidence Interval:
$\$3,494.80 < \mu_1 - \mu_2 < \$8,021.20$

**7.**

| Maximum | Minimum | D | D$^2$ |
|---|---|---|---|
| 44 | 27 | 17 | 289 |
| 46 | 34 | 12 | 144 |
| 46 | 24 | 22 | 484 |
| 36 | 19 | 17 | 289 |
| 34 | 19 | 15 | 225 |
| 36 | 26 | 10 | 100 |
| 57 | 33 | 24 | 576 |
| 62 | 57 | 5 | 25 |
| 73 | 46 | 27 | 729 |
| 53 | 26 | 27 | 729 |
| | | $\sum D = 176$ | $\sum D^2 = 3590$ |

$H_0$: $\mu_D = 10$
$H_1$: $\mu_D > 10$

C. V. = 2.821
$\bar{D} = \frac{176}{10} = 17.6$

$$s_D = \sqrt{\frac{3590 - \frac{176^2}{10}}{9}} = 7.3967$$

$$t = \frac{17.6 - 10}{\frac{7.3967}{\sqrt{10}}} = 3.249$$

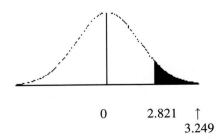

$$0 \qquad 2.821 \quad \uparrow$$
$$3.249$$

Reject the null hypothesis. There is enough evidence to support the claim that the mean difference is more than 10 degrees.

**9.**
$\hat{p}_1 = \frac{207}{365} = 0.567 \qquad \hat{p}_2 = \frac{166}{365} = 0.455$

$\bar{p} = \frac{207 + 166}{365 + 365} = 0.511$

$\bar{q} = 1 - 0.51 = 0.489$

$H_0$: $p_1 = p_2$
$H_1$: $p_1 \neq p_2$ (claim)

C. V. = $\pm 2.33$

$$z = \frac{(0.567 - 0.455) - 0}{\sqrt{(0.511)(0.489)(\frac{1}{365} + \frac{1}{365})}} = 3.03$$

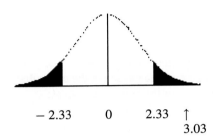

$$-2.33 \qquad 0 \qquad 2.33 \quad \uparrow$$
$$3.03$$

Reject the null hypothesis. There is enough evidence to support the claim that the proportions are different.

For the 98% confidence interval:
$$(\hat{p}_1 - \hat{p}_2) - z_{\frac{\alpha}{2}}\sqrt{\frac{\hat{p}_1\hat{q}_1}{n_1} + \frac{\hat{p}_2\hat{q}_2}{n_2}} < p_1 - p_2 <$$
$$(\hat{p}_1 - \hat{p}_2) + z_{\frac{\alpha}{2}}\sqrt{\frac{\hat{p}_1\hat{q}_1}{n_1} + \frac{\hat{p}_2\hat{q}_2}{n_2}}$$

$\hat{p}_1 = 0.567 \qquad \hat{q}_1 = 0.433$

$\hat{p}_2 = 0.455 \qquad \hat{q}_2 = 0.545$

$$(0.567 - 0.455) - 2.33\sqrt{\frac{(0.567)(0.433)}{365} + \frac{(0.455)(0.545)}{365}}$$

$$< p_1 - p_2 < (0.567 - 0.455) + 2.33\sqrt{\frac{(0.567)(0.433)}{365} + \frac{(0.455)(0.545)}{365}}$$

$0.112 - 0.086 < p_1 - p_2 < 0.112 + 0.086$

$0.026 < p_1 - p_2 < 0.198$

**11.**
$H_0$: $\sigma_1 = \sigma_2$
$H_1$: $\sigma_1 \neq \sigma_2$ (claim)

$\alpha = 0.10$
dfN = 23 \qquad dfD = 10
C. V. = 2.77

11. continued

$$F = \frac{13.2^2}{4.1^2} = 10.365$$

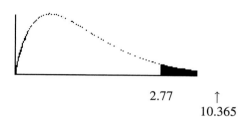

2.77          ↑
         10.365

Reject the null hypothesis. There is enough evidence to support the claim that there is a difference in the standard deviations.

CHAPTER 9 QUIZ

1. False, there are different formulas for independent and dependent samples.
2. False, the samples are independent.
3. True
4. False, they can be right, left, or two tailed.
5. d.
6. a.
7. c.
8. b.
9. $\mu_1 = \mu_2$
10. pooled
11. normal
12. negative
13. $\frac{s_1^2}{s_2^2}$

14. $H_0$: $\mu_1 = \mu_2$
$H_1$: $\mu_1 \neq \mu_2$   (claim)
C. V. $= \pm 2.58$   z $= -3.69$
Reject the null hypothesis. There is enough evidence to support the claim that there is a difference in the cholesterol levels of the two groups.
99% Confidence Interval:
$-10.2 < \mu_1 - \mu_2 < -1.8$

15. $H_0$: $\mu_1 = \mu_2$
$H_1$: $\mu_1 > \mu_2$   (claim)
C. V. $= 1.28$       z $= 1.60$
Reject the null hypothesis. There is enough evidence to support the claim that average rental fees for the east apartments is greater than the average rental fees for the west apartments.

16. $H_0$: $\mu_1 = \mu_2$
$H_1$: $\mu_1 \neq \mu_2$   (claim)
t $= 10.922$        C. V. $= \pm 2.779$
Reject the null hypothesis. There is enough evidence to support the claim that the average prices are different.

$0.298 < \mu_1 - \mu_2 < 0.502$

17. $H_0$: $\mu_1 = \mu_2$
$H_1$: $\mu_1 < \mu_2$   (claim)
C. V. $= -1.860$ d.f. $= 8$  t $= -4.05$
Reject the null hypothesis. There is enough evidence to support the claim that accidents have increased.

18. $H_0$: $\mu_1 = \mu_2$
$H_1$ $\mu_1 \neq \mu_2$   (claim)
C. V. $= \pm 2.718$ t $= 9.807$
Reject the null hypothesis. There is enough evidence to support the claim that the salaries are different.

$\$6653 < \mu_1 - \mu_2 < \$11,757$

19. $H_0$: $\mu_1 = \mu_2$
$H_1$ $\mu_1 > \mu_2$   (claim)
d. f. $= 10$        t $= 0.874$
$0.10 < $ P-value $< 0.25$ (0.198)
Do not reject the null hypothesis since P-value $> 0.05$. There is not enough evidence to support the claim that the incomes of city residents are greater than the incomes of rural residents.

20. $H_0$: $\mu_D = 0$
$H_1$ $\mu_D < 0$   (claim)
C. V. $= -2.821$ t $= -2.44$
Do not reject the null hypothesis. There is not enough evidence to support the claim that the sessions improved math skills.

21. $H_0$: $\mu_D = 0$
$H_1$ $\mu_D < 0$   (claim)
C. V. $= -1.833$ t $= -2.02$
Reject the null hypothesis. There is enough evidence to support the claim that egg production was increased.

22. $H_0$: $p_1 = p_2$
$H_1$: $p_1 \neq p_2$   (claim)
C. V. $= \pm 1.65$   z $= -0.69$

22. continued
Do not reject the null hypothesis. There is
not enough evidence to support the claim
that the proportions are different.

90% Confidence Interval:
$-0.105 < p_1 - p_2 < 0.045$

23. $H_0$: $p_1 = p_2$  (claim)
$H_1$: $p_1 \neq p_2$
C. V. $= \pm 1.96$  $z = 0.544$
Do not reject the null hypothesis. There is
not enough evidence to support the claim
that the proportions have changed.

95% Confidence Interval:
$-0.026 < p_1 - p_2 < 0.0460$

Yes, the confidence interval contains 0;
hence, the null hypothesis is not rejected.

24. $H_0$: $\sigma_1^2 = \sigma_2^2$
$H_1$ $\sigma_1^2 \neq \sigma_2^2$  (claim)
d. f. N. $= 17$      d. f. D. $= 14$
$F = 1.637$
P-value $> 0.20$ (0.357)
Do not reject since P-value $> 0.05$. There is
not enough evidence to support the claim
that the variances are different.

25. $H_0$: $\sigma_1^2 = \sigma_2^2$
$H_1$ $\sigma_1^2 \neq \sigma_2^2$
C. V. $= 1.90$      $F = 1.296$
Do not reject. There is not enough evidence
to support the claim that the variances are
different.

Note: Graphs are not to scale and are intended to convey a general idea.

Answers may vary due to rounding, TI-83's, or computer programs.

EXERCISE SET 10-1

1.
Two variables are related when there exists a discernible pattern between them.

3.
$r$, $\rho$ (rho)

5.
A positive relationship means that as x increases, y also increases.
A negative relationship means that as x increases, y decreases.

7.
Answers will vary.

9.
Pearson's Product Moment Correlation Coefficient.

11.
There are many other possibilities, such as chance, relationship to a third variable, etc.

13.

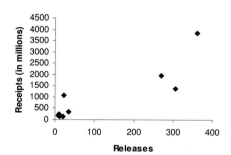

$\sum x = 1045$
$\sum y = 9283$
$\sum x^2 = 299{,}315$
$\sum y^2 = 21{,}881{,}839$
$\sum xy = 2{,}380{,}435$
n = 9
$r = \dfrac{n(\sum xy)-(\sum x)(\sum y)}{\sqrt{[n(\sum x^2)-(\sum x)^2]\,[n(\sum y^2)-(\sum y)^2]}}$

$r = \dfrac{9(2{,}380{,}435)-(1045)(9283)}{\sqrt{[9(299{,}315)-(1045)^2]\,[9(21{,}881{,}839)-(9283)^2]}}$

13. continued
$r = 0.880$

$H_0$: $\rho = 0$
$H_1$: $\rho \neq 0$
C. V. = $\pm 0.666$     d. f. = 7

Decision: Reject.  There is a significant linear relationship between number of movie releases and gross receipts.

15.

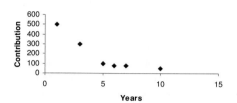

$\sum x = 32$
$\sum y = 1105$
$\sum x^2 = 220$
$\sum y^2 = 364{,}525$
$\sum xy = 3405$
n = 6
$r = \dfrac{n(\sum xy)-(\sum x)(\sum y)}{\sqrt{[n(\sum x^2)-(\sum x)^2]\,[n(\sum y^2)-(\sum y)^2]}}$

$r = \dfrac{6(3405)-(32)(1105)}{\sqrt{[6(220)-(32)^2][6(364525)-(1105)^2]}}$

$r = -0.883$

$H_0$: $\rho = 0$
$H_1$: $\rho \neq 0$
C. V. = $\pm 0.811$     d. f. = 4
Decision:  Reject.  There is a significant linear relationship between the number of years a person has been out of school and his or her contributions.

17.

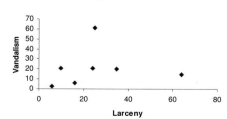

101

17. continued

$\sum x = 180$
$\sum y = 147$
$\sum x^2 = 6914$
$\sum y^2 = 5273$
$\sum xy = 4013$
$n = 7$

$$r = \frac{n(\sum xy) - (\sum x)(\sum y)}{\sqrt{[n(\sum x^2) - (\sum x)^2][n(\sum y^2) - (\sum y)^2]}}$$

$$r = \frac{7(4013) - (180)(147)}{\sqrt{[7(6914) - (180)^2][7(5273) - (147)^2]}}$$

$r = 0.104$

$H_0$: $\rho = 0$
$H_1$: $\rho \neq 0$
C. V. $= \pm 0.754$    d. f. $= 5$
Decision: Do not reject. There is no significant linear relationship between the number of larceny crimes and the number of vandalism crimes committed on college campuses in southwestern Pennsylvania.

19.

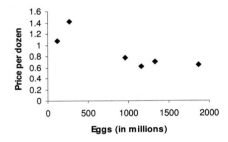

$\sum x = 5709$
$\sum y = 5.236$
$\sum x^2 = 7,609,557$
$\sum y^2 = 5.067302$
$\sum xy = 4115.025$
$n = 6$

$$r = \frac{6(4115.025) - (5709)(5.236)}{\sqrt{[6(7,609,557) - (5709)^2][6(5.067302) - (5.236)^2]}}$$

$r = -0.833$

$H_0$: $\rho = 0$
$H_1$: $\rho \neq 0$
C. V. $= \pm 0.811$    d. f. $= 4$
Decision: Reject. There is a significant linear relationship between the number of eggs produced and price per dozen.

21.

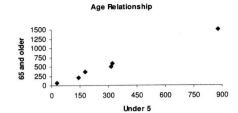

$\sum x = 1862$
$\sum y = 3222$
$\sum x^2 = 1,026,026$
$\sum y^2 = 3,009,596$
$\sum xy = 1,754,975$
$n = 6$

$$r = \frac{n(\sum xy) - (\sum x)(\sum y)}{\sqrt{[n(\sum x^2) - (\sum x)^2][n(\sum y^2) - (\sum y)^2]}}$$

$$r = \frac{6(1,754,975) - (1862)(3222)}{\sqrt{[6(1,026,026) - 1862^2][6(3,009,596) - 3222^2]}}$$

$r = 0.997$

$H_0$: $\rho = 0$
$H_1$: $\rho \neq 0$
C. V. $= \pm 0.811$        d. f. $= 4$
Decision: Reject. There is a significant linear relationship between the under 5 age group and the 65 and over age group.

23.

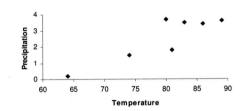

$\sum x = 557$
$\sum y = 17.7$
$\sum x^2 = 44,739$
$\sum y^2 = 55.99$
$\sum xy = 1468.9$
$n = 7$

$$r = \frac{n(\sum xy) - (\sum x)(\sum y)}{\sqrt{[n(\sum x^2) - (\sum x)^2][n(\sum y^2) - (\sum y)^2]}}$$

$$r = \frac{7(1468.9) - (557)(17.7)}{\sqrt{[7(44,739) - 557^2][7(55.99) - 17.7^2]}}$$

$r = 0.883$

23. continued

$H_0$: $\rho = 0$

$H_1$: $\rho \neq 0$

C. V. $= \pm 0.754$     d. f. $= 5$

Decision: Reject. There is a significant linear relationship between the average daily temperature and the average monthly precipitation.

25.

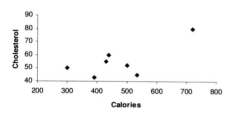

**Calories vs. Cholesterol**

$\sum x = 3315$

$\sum y = 385$

$\sum x^2 = 1,675,225$

$\sum y^2 = 22,103$

$\sum xy = 189,495$

n = 7

$$r = \frac{n(\sum xy)-(\sum x)(\sum y)}{\sqrt{[n(\sum x^2)-(\sum x)^2]\,[n(\sum y^2)-(\sum y)^2]}}$$

$$r = \frac{7(189,495)-(3315)(385)}{\sqrt{[7(1,675,225)-(3315)^2][7(22,103)-(385)^2]}}$$

$r = 0.725$

$H_0$: $\rho = 0$

$H_1$: $\rho \neq 0$

C. V. $= \pm 0.754$     d. f. $= 5$

Decision: Do not reject. There is a no significant linear relationship between the number of calories and the cholesterol content of fast-food chicken sandwiches.

27.

**Licensed Beds vs. Staffed Beds**

$\sum x = 1013$

$\sum y = 748$

$\sum x^2 = 168,435$

27. continued

$\sum y^2 = 90,626$

$\sum xy = 120,953$

n = 7

$$r = \frac{7(120,953)-(1013)(748)}{\sqrt{[7(168,435)-(1013)^2]\,[7(90,626)-(748)^2]}}$$

$r = 0.831$

$H_0$: $\rho = 0$

$H_1$: $\rho \neq 0$

C. V. $= \pm 0.754$    d. f. $= 5$

Decision: Reject. There is a significant linear relationship between the number of licensed beds in a hospital and the number of staffed beds.

29.

$$r = \frac{n(\sum xy)-(\sum x)(\sum y)}{\sqrt{[n(\sum x^2)-(\sum x)^2]\,[n(\sum y^2)-(\sum y)^2]}}$$

$$r = \frac{5(125)-(15)(35)}{\sqrt{[5(55)-(15)^2][5(285)-(35)^2]}} = 1$$

$$r = \frac{5(125)-(35)(15)}{\sqrt{[5(285)-(35)^2][5(55)-(15)^2]}} = 1$$

The value of $r$ does not change when the values for x and y are interchanged.

**EXERCISE 10-2**

1.

Draw the scatter plot and test the significance of the correlation coefficient.

3.

$y' = a + bx$

5.

It is the line that is drawn through the points on the scatter plot such that the sum of the squares of the vertical distances each point is from the line is at a minimum.

7.

When $r$ is positive, $b$ will be positive. When $r$ is negative, $b$ will be negative.

9.

The closer $r$ is to $+1$ or $-1$, the more accurate the predicted value will be.

11.

When $r$ is not significant, the mean of the $y$ values should be used to predict $y$.

13.

$$a = \frac{(\sum y)(\sum x^2) - (\sum x)(\sum xy)}{n(\sum x^2) - (\sum x)^2}$$

$$a = \frac{(9283)(299,315) - (1045)(2,380,435)}{9(299,315) - (1045)^2} = 181.661$$

$$b = \frac{n(\sum xy) - (\sum x)(\sum y)}{n(\sum x^2) - (\sum x)^2}$$

$$b = \frac{9(2,380,435) - (1045)(9283)}{9(299,315) - (1045)^2} = 7.319$$

$y' = a + bx$
$y' = 181.661 + 7.319x$
$y' = 181.661 + 7.319(200) = \$1645.5$
million gross receipts

15.

$$a = \frac{(\sum y)(\sum x^2) - (\sum x)(\sum xy)}{n(\sum x^2) - (\sum x)^2}$$

$$a = \frac{(1105)(220) - (32)(3405)}{6(220) - (32)^2}$$

$$a = \frac{243100 - 108960}{1320 - 1024} = \frac{134140}{296} = 453.176$$

$$b = \frac{n(\sum xy) - (\sum x)(\sum y)}{n(\sum x^2) - (\sum x)^2}$$

$$b = \frac{6(3405) - (32)(1105)}{6(220) - (32)^2} = \frac{20430 - 35360}{296}$$

$$b = \frac{-14930}{296} = -50.439$$

$y' = a + bx$
$y' = 453.176 - 50.439x$
$y' = 453.176 - 50.439(4) = \$251.42$

17.
Since $r$ is not significant, no regression should be done.

19.

$$a = \frac{(\sum y)(\sum x^2) - (\sum x)(\sum xy)}{n(\sum x^2) - (\sum x)^2}$$

$$a = \frac{(5.236)(7,609,557) - (5709)(4115.025)}{6(7,609,557) - (5709)^2}$$

$a = 1.252$

$$b = \frac{n(\sum xy) - (\sum x)(\sum y)}{n(\sum x^2) - (\sum x)^2}$$

$$b = \frac{6(4115.025) - (5709)(5.236)}{6(4115.025) - (5709)^2}$$

$b = -0.000398$

$y' = a + bx$
$y' = 1.252 - 0.000398x$

19. continued
$y' = 1.252 - 0.000398(1600) = 0.615$
per dozen

21.

$$a = \frac{(\sum y)(\sum x^2) - (\sum x)(\sum xy)}{n(\sum x^2) - (\sum x)^2}$$

$$a = \frac{(3222)(1026026) - (1862)(1754975)}{6(1026026) - (1862)^2}$$

$a = 14.165$

$$b = \frac{n(\sum xy) - (\sum x)(\sum y)}{n(\sum x^2) - (\sum x)^2}$$

$$b = \frac{6(1754975) - (1862)(3222)}{6(1026026) - (1862)^2} = 1.685$$

$y' = a + bx$
$y' = 14.165 + 1.685x$
$y' = 14.165 + 1.685(200) = 351$ under 5.

23.

$$a = \frac{(\sum y)(\sum x^2) - (\sum x)(\sum xy)}{n(\sum x^2) - (\sum x)^2}$$

$$a = \frac{(17.7)(44739) - (557)(1468.9)}{7(44739) - (557)^2} = -8.994$$

$$b = \frac{n(\sum xy) - (\sum x)(\sum y)}{n(\sum x^2) - (\sum x)^2}$$

$$b = \frac{7(1468.9) - (557)(17.7)}{7(44739) - (557)^2} = 0.1448$$

$y' = a + bx$
$y' = -8.994 + 0.1448x$
$y' = -8.994 + 0.1448(70) = 1.1$ inches

25.
Since $r$ is not significant, no regression should be done.

27.

$$a = \frac{(\sum y)(\sum x^2) - (\sum x)(\sum xy)}{n(\sum x^2) - (\sum x)^2}$$

$$a = \frac{(748)(168,435) - (1013)(120,953)}{7(168,435) - (1013)^2}$$
$a = 22.659$

$$b = \frac{n(\sum xy) - (\sum x)(\sum y)}{n(\sum x^2) - (\sum x)^2}$$

$$b = \frac{7(120,953) - (1013)(748)}{7(168,435) - (1013)^2} = 0.582$$

$y' = a + bx$
$y' = 22.659 + 0.582x$
$y' = 22.659 + 0.582(44) = 48.267$ staffed beds

29.

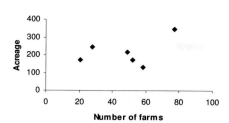

$\sum x = 285$
$\sum y = 1289$
$\sum x^2 = 15,637.88$
$\sum y^2 = 305,731$
$\sum xy = 64,565.8$
n = 6

$$r = \frac{n(\sum xy)-(\sum x)(\sum y)}{\sqrt{[n(\sum x^2)-(\sum x)^2][n(\sum y^2)-(\sum y)^2]}}$$

$$r = \frac{6(64,565.8)-(285)(1289)}{\sqrt{[6(15,637.88)-(285)^2][6(305,731)-(1289)^2]}}$$

$r = 0.429$

$H_0$: $\rho = 0$
$H_1$: $\rho \neq 0$
C. V. = $\pm 0.811$    d. f. = 4

Decision: Do not reject
There is no significant relationship between the number of farms and acreage.

31.

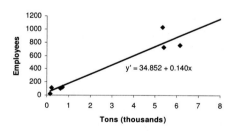

Tons of Coal vs. Number of Employees

$\sum x = 26,728$
$\sum y = 4027$
$\sum x^2 = 162,101,162$
$\sum y^2 = 3,550,103$
$\sum xy = 23,663,669$
n = 8

$$r = \frac{n(\sum xy)-(\sum x)(\sum y)}{\sqrt{[n(\sum x^2)-(\sum x)^2][n(\sum y^2)-(\sum y)^2]}}$$

$$r = \frac{8(23662669)-(26728)(4027)}{\sqrt{[8(162101162)-26728^2][8(3550103)-(4027)^2]}}$$

31. continued
$r = 0.970$

$H_0$: $\rho = 0$
$H_1$: $\rho \neq 0$
C. V. = $\pm 0.707$    d. f. = 6
Decision: Reject. There is a significant relationship between the number of tons of coal produced and the number of employees.

$$a = \frac{(\sum y)(\sum x^2)-(\sum x)(\sum xy)}{n(\sum x^2)-(\sum x)^2}$$

$$a = \frac{(4027)(162101162)-(26728)(23663669)}{8(162101162)-(26728)^2}$$

$a = 34.852$

$$b = \frac{n(\sum xy)-(\sum x)(\sum y)}{n(\sum x^2)-(\sum x)^2}$$

$$b = \frac{8(23663669)-(26728)(4027)}{8(162101162)-(26728)^2} = 0.140$$

$y' = a + bx$
$y' = 34.852 + 0.140x$
$y' = 34.852 + 0.140(500) = 104.9$

33.

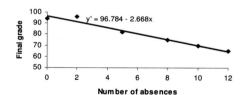

Absences vs. Final Grades

$\sum x = 37$
$\sum y = 482$
$\sum x^2 = 337$
$\sum y^2 = 39526$
$\sum xy = 2682$
n = 6

$$r = \frac{n(\sum xy)-(\sum x)(\sum y)}{\sqrt{[n(\sum x^2)-(\sum x)^2][n(\sum y^2)-(\sum y)^2]}}$$

$$r = \frac{6(2682)-(37)(482)}{\sqrt{[6(337)-(37)^2][6(39526)-(482)^2]}}$$

$r = -0.981$

$H_0$: $\rho = 0$
$H_1$: $\rho \neq 0$
C. V. = $\pm 0.811$    d. f. = 4

**33. continued**
Decision: Reject. There is a significant relationship between the number of absences and the final grade.

$$a = \frac{(\sum y)(\sum x^2)-(\sum x)(\sum xy)}{n(\sum x^2)-(\sum x)^2}$$

$$a = \frac{(482)(337)-(37)(2682)}{6(337)-(37)^2} = 96.784$$

$$b = \frac{n(\sum xy)-(\sum x)(\sum y)}{n(\sum x^2)-(\sum x)^2}$$

$$b = \frac{6(2682)-(37)(482)}{6(337)-(37)^2} = -2.668$$

$y' = a + bx$
$y' = 96.784 - 2.668x$

**35.**

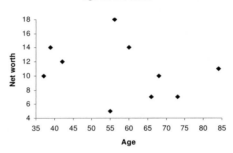

Age vs. Net Worth

$\sum x = 580$
$\sum y = 108$
$\sum x^2 = 35{,}780$
$\sum y^2 = 1304$
$\sum xy = 6120$
$n = 10$

$$r = \frac{n(\sum xy)-(\sum x)(\sum y)}{\sqrt{[n(\sum x^2)-(\sum x)^2]\,[n(\sum y^2)-(\sum y)^2]}}$$

$$r = \frac{10(6120)-(580)(108)}{\sqrt{10(35{,}780)-580^2][10(1304)-108^2]}}$$

$r = -0.265$
$H_0$: $\rho = 0$
$H_1$: $\rho \neq 0$
P-value > 0.05 (0.459)
Decision: Do not reject since P-value > 0.05. There is no significant linear relationship between the ages of billionaires and their net worth. Since $r$ is not significant, no regression should be done.

**37.**
For Exercise 15:
$\bar{x} = 5.3333$

**37. continued**
$\bar{y} = 184.1667$
$b = -50.439$
$a = \bar{y} - b\bar{x}$
$a = 184.1667 - (-50.439)(5.3333)$
$a = 184.1667 + 269.0063$
$a = 453.173$ (differs due to rounding)

For Exercise 16:
Since $r$ is not significant, no regression should be done.

**EXERCISE SET 10-3**

**1.**
Explained variation is the variation due to the relationship and is computed by $\sum(y' - \bar{y})^2$.

**3.**
Total variation is the sum of the squares of the vertical distances of the points from the mean. It is computed by $\sum(y - \bar{y})^2$.

**5.**
It is found by squaring $r$.

**7.**
The coefficient of non-determination is $1 - r^2$.

**9.**
For $r = 0.75$, $r^2 = 0.5625$ and $1 - r^2 = 0.4375$.
Thus 56.25% of the variation of y is due to the variation of x, and 43.75% of the variation of y is due to chance.

**11.**
For $r = 0.42$, $r^2 = 0.1764$ and $1 - r^2 = 0.8236$. Thus 17.64% of the variation of y is due to the variation of x, and 82.36% of the variation of y is due to chance.

**13.**
For $r = 0.91$, $r^2 = 0.8281$ and $1 - r^2 = 0.1719$. Thus 82.81% of the variation of y is due to the variation of x, and 17.19% of the variation of y is due to chance.

Note: For Exercises 15 − 17, values for $a$ and $b$ are rounded to 3 decimal places according to the textbook's rounding rule for intercept and slope of the regression equation. Where these answers differ from the text, additional decimal places are included to show consistency with text answers.

15.
$$S_{est} = \sqrt{\frac{\sum y^2 - a\sum y - b\sum xy}{n-2}}$$

$$S_{est} = \sqrt{\frac{21,881,839 - 181.661(9283) - (7.319)(2,380,435)}{9-2}}$$

$$S_{est} = \sqrt{396,153.7389} = 629.41$$

Using $a = 181.661102$ and $b = 7.318708$,
$S_{est} = 629.4862$

17.
$$S_{est} = \sqrt{\frac{\sum y^2 - a\sum y - b\sum xy}{n-2}} =$$

$$S_{est} = \sqrt{\frac{364525 - (453.176)(1105) - (-50.439)(3405)}{6-2}}$$

$$S_{est} = 94.22$$

19.
$y' = 181.661 + 7.319x$
$y' = 181.661 + 7.319(200)$
$y' = 1645.461$

$$y' - t_{\frac{\alpha}{2}} \cdot S_{est}\sqrt{1 + \frac{1}{n} + \frac{n(x-\overline{X})}{n\sum x^2 - (\sum x)^2}} < y <$$

$$y' + t_{\frac{\alpha}{2}} \cdot S_{est}\sqrt{1 + \frac{1}{n} + \frac{n(x-\overline{X})^2}{n\sum x^2 - (\sum x)^2}}$$

$$1645.461 - (1.895)(629.4862)\sqrt{1 + \frac{1}{9} + \frac{9(200-116.11)^2}{9(299,315)-1045^2}}$$

$$< y < 1645.461 +$$

$$(1.895)(629.4862)\sqrt{1 + \frac{1}{9} + \frac{9(200-116.11)^2}{9(299,315)-1045^2}}$$

$$1645.461 - 1279.580227 < y <$$

$$1645.461 + 1279.580227$$

$$365.88 < y < 2925.04$$

21.
$y' = 453.176 - 50.439x$

21. continued
$y' = 453.176 - 50.439(4)$
$y' = 251.42$

$$y' - t_{\frac{\alpha}{2}} \cdot S_{est}\sqrt{1 + \frac{1}{n} + \frac{n(x-\overline{X})^2}{n\sum x^2 - (\sum x)^2}} < y <$$

$$y' + t_{\frac{\alpha}{2}} \cdot S_{est}\sqrt{1 + \frac{1}{n} + \frac{n(x-\overline{X})^2}{n\sum x^2 - (\sum x)^2}}$$

$$251.42 - 2.132(94.22)\sqrt{1 + \frac{1}{6} + \frac{6(4-5.33)^2}{6(220)-32^2}}$$

$$< y < 251.42 + 2.132(94.22)\sqrt{1 + \frac{1}{6} + \frac{6(4-5.33)^2}{6(220)-32^2}}$$

$$251.42 - (2.132)(94.22)(1.1) < y < 251.42 + (2.132)(94.22)(1.1)$$
$$\$30.46 < y < \$472.38$$

EXERCISE SET 10-4

1.
Simple linear regression has one independent variable and one dependent variable. Multiple regression has one dependent variable and two or more independent variables.

3.
The relationship would include all variables.

5.
The multiple correlation coefficient R is always higher than the individual correlation coefficients. Also, the value of R can range from 0 to +1.

7.
$y' = 9.6 + 2.2x_1 - 1.08x_2$
$y' = 9.6 + 2.2(9) - 1.08(24) = 3.48$

9.
$y' = -14.9 + 0.93359x_1 + 0.99847x_2 + 5.3844x_3$
$y' = -14.9 + 0.93359(8) + 0.99847(34) + 5.3844(11)$
$y' = 85.75$ (grade) or 86

11.
R is a measure of the strength of the relationship between the dependent variables and all the independent variables.

13.

$R^2$ is the coefficient of multiple determination. $R^2_{adj}$ is adjusted for sample size and the number of predictors.

15.

The F test is used to test the significance of R.

REVIEW EXERCISES - CHAPTER 10

1.

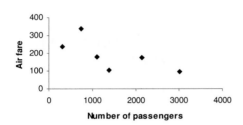

$\sum x = 8722$
$\sum y = 1132$
$\sum x^2 = 17,545,122$
$\sum y^2 = 254,234$
$\sum xy = 1,340,270$
n = 6

$r = \frac{n(\sum xy)-(\sum x)(\sum y)}{\sqrt{[n(\sum x^2)-(\sum x)^2][n(\sum y^2)-(\sum y)^2]}}$

$r = \frac{6(1,340,270)-(8722)(1132)}{\sqrt{[6(17,545,122)-(8722)^2][6(254,234)-(1132)^2]}}$

$r = -0.686$

$H_0$: $\rho = 0$
$H_1$: $\rho \neq 0$
C. V. $= \pm 0.917$ at $\alpha = 0.01$    d. f. $= 4$

Decision: Do not reject. There is no significant linear relationship between the number of passengers and airfare.

3.

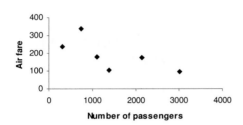

3. continued
$\sum x = 171.1$
$\sum y = 11.21$
$\sum x^2 = 4406.73$
$\sum y^2 = 19.9107$
$\sum xy = 293.925$
n = 7

$r = \frac{n(\sum xy)-(\sum x)(\sum y)}{\sqrt{[n(\sum x^2)-(\sum x)^2] [n(\sum y^2)-(\sum y)^2]}}$

$r = \frac{7(293.925)-(171.1)(11.21)}{\sqrt{[7(4406.73)-(171.1)^2][7(19.9107)-(11.21)^2]}}$

$r = 0.95$

$H_0$: $\rho = 0$
$H_1$: $\rho \neq 0$
C. V. $= \pm 0.875$    d. f. $= 5$

Decision: Reject. There is a significant relationship between gas tax and cigarette tax.

$a = \frac{(\sum y)(\sum x^2)-(\sum x)(\sum xy)}{n(\sum x^2)-(\sum x)^2}$

$a = \frac{(11.21)(4406.73)-(171.1)(293.925)}{7(4406.73)-(171.1)^2} = -0.567$

$b = \frac{n(\sum xy)-(\sum x)(\sum y)}{n(\sum x^2)-(\sum x)^2}$

$b = \frac{7(293.925)-(171.1)(11.21)}{7(4406.73)-(171.1)^2} = 0.089$

$y' = a + bx$
$y' = -0.567 + 0.089x$
$y' = -0.567 + 0.089(18.4) = \$1.07$

5.

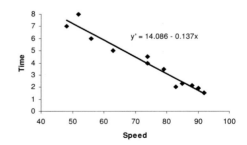

$\sum x = 884$
$\sum y = 47.8$
$\sum x^2 = 67,728$
$\sum y^2 = 242.06$
$\sum xy = 3163.8$
n = 12

5. continued

$$r = \frac{n(\sum xy)-(\sum x)(\sum y)}{\sqrt{[n(\sum x^2)-(\sum x)^2][n(\sum y^2)-(\sum y)^2]}}$$

$$r = \frac{12(3163.8)-(884)(47.8)}{\sqrt{[12(67728)-(884)^2][12(242.06)-(47.8)^2]}}$$

$$r = -0.974$$

$H_0$: $\rho = 0$
$H_1$: $\rho \neq 0$
C. V. $= \pm 0.708$    d. f. $= 10$

Decision: Reject. There is a significant relationship between speed and time.

$$a = \frac{(\sum y)(\sum x^2)-(\sum x)(\sum xy)}{n(\sum x^2)-(\sum x)^2}$$

$$a = \frac{(47.8)(67728)-(884)(3163.8)}{12(67728)-(884)^2}$$

$$a = 14.086$$

$$b = \frac{n(\sum xy)-(\sum x)(\sum y)}{n(\sum x^2)-(\sum x)^2}$$

$$b = \frac{12(3163.8)-(884)(47.8)}{12(67728)-(884)^2}$$

$$b = -0.137$$

$y' = a + bx$
$y' = 14.086 - 0.137x$
$y' = 14.086 - 0.137(72) = 4.222$

7.

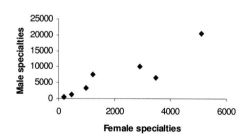

$\sum x = 14,301$
$\sum y = 49,452$
$\sum x^2 = 49,158,943$
$\sum y^2 = 630,840,804$
$\sum xy = 168,984,784$
$n = 7$
$$r = \frac{n(\sum xy)-(\sum x)(\sum y)}{\sqrt{[n(\sum x^2)-(\sum x)^2][n(\sum y^2)-(\sum y)^2]}}$$

7. continued

$$r = \frac{7(168,984,784)-(14,301)(49,452)}{\sqrt{[7(49,158,943)-(14,301)^2][7(630,840,804)-(49,452)^2]}}$$

$$r = 0.907$$

$H_0$: $\rho = 0$
$H_1$: $\rho \neq 0$
C. V. $\pm 0.875$    d. f. $= 5$

Decision: Reject. There is a significant relationship between female specialties and male specialties.

$$a = \frac{(\sum y)(\sum x^2)-(\sum x)(\sum xy)}{n(\sum x^2)-(\sum x)^2}$$

$$a = \frac{(49,452)(49,158,943)-(14,301)(168,984,784)}{7(49,158,943)-(14,301)^2}$$

$$a = 102.846$$

$$b = \frac{n(\sum xy)-(\sum x)(\sum y)}{n(\sum x^2)-(\sum x)^2}$$

$$b = \frac{7(168,984,784)-(14,301)(49,452)}{7(49,158,943)-(14,301)^2}$$

$$b = 3.408$$

$y' = a + bx$
$y' = 102.846 + 3.408x$
$y' = 102.846 + 3.408(2000)$
$y' = 6918.8$ or $6919$

9.
$$S_{est} = \sqrt{\frac{\sum y^2 - a\sum y - b\sum xy}{n-2}}$$

$$S_{est} = \sqrt{\frac{242.06 - 14.086(47.8) + 0.137(3163.8)}{12-2}}$$

$$S_{est} = \sqrt{\frac{2.1898}{10}} = \sqrt{0.21898} = 0.468$$

(Note: TI-83 calculator answer is 0.513)

11.
(For calculation purposes only, since no regression should be done.)

$y' = 14.086 - 0.137x$
$y' = 14.086 - 0.137(72) = 4.222$

$$y' - t_{\frac{\alpha}{2}} \cdot S_{est}\sqrt{1 + \frac{1}{n} + \frac{n(x-\overline{X})^2}{n\Sigma x^2 - (\Sigma x)^2}} < y <$$

$$y' + t_{\frac{\alpha}{2}} \cdot S_{est}\sqrt{1 + \frac{1}{n} + \frac{n(x-\overline{X})^2}{n\Sigma x^2 - (\Sigma x)^2}}$$

11. continued

$$4.222 - 1.812(0.468)\sqrt{1 + \tfrac{1}{12} + \tfrac{12(72-73.667)^2}{12(67,728)-884^2}}$$
$$< y < 4.222 + 1.812(0.468)\sqrt{1 + \tfrac{1}{12} + \tfrac{12(72-73.667)^2}{12(67,728)-884^2}}$$

$4.222 - 1.812(0.468)(1.041) < y <$
    $4.222 + 1.812(0.468)(1.041)$
$3.34 < y < 5.10$

13.
$y' = 12.8 + 2.09X_1 + 0.423X_2$
$y' = 12.8 + 2.09(4) + 0.423(2) = 22.006$ or
22.01

CHAPTER 10 QUIZ

1. False, the y variable would decrease.
2. True
3. True
4. False, the relationship may be affected by another variable, or by chance.
5. False, a relationship may be caused by chance.
6. False, there are several independent variables and one dependent variable.
7. a.
8. a.
9. d.
10. c.
11. b.
12. scatter plot
13. independent
14. $-1, +1$
15. b.
16. line of best fit
17. $+1, -1$

18.

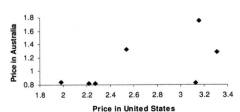

Price Comparison of Drugs

$\sum x = 18.61$
$\sum x^2 = 51.1919$
$\sum y = 7.67$
$\sum y^2 = 9.2083$
$\sum xy = 21.0956$
$n = 7$
$r = 0.600$

18. continued
$H_0: \rho = 0$
$H_1: \rho \neq 0$
C.V. $= \pm 0.754$    d. f. $= 5$
Do not reject. There is no significant linear relationship between the price of the same drugs in the United States and in Australia. No regression should be done.

19.

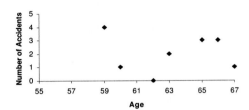

Age vs. Number of Accidents

$\sum x = 442$
$\sum x^2 = 27,964$
$\sum y = 14$
$\sum y^2 = 40$
$\sum xy = 882$
$n = 7$
$r = -0.078$
$H_0: \rho = 0$
$H_1: \rho \neq 0$
C. V. $= \pm 0.754$    d. f. $= 5$
Decision: do not reject. There is not a significant relationship between age and number of accidents. No regression should be done.

20.

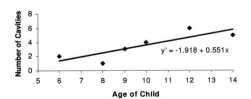

Age vs. Number of Cavities

$\sum x = 59$
$\sum x^2 = 621$
$\sum y = 21$
$\sum y^2 = 91$
$\sum xy = 229$
$n = 6$
$r = 0.842$
$H_0: \rho = 0$
$H_1: \rho \neq 0$
C. V. $= \pm 0.811$    d. f. $= 4$

**20. continued**
Decision: Reject. There is a significant linear relationship between age and number of cavities.

$a = -1.918 \qquad b = 0.551$
$y' = -1.918 + 0.551x$
When x = 11: $y' = -1.918 + 0.551(11)$
$y' = 4.14$ or 4 cavities

**21.**

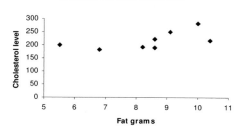

$\sum x = 67.2$
$\sum x^2 = 582.62$
$\sum y = 1740$
$\sum y^2 = 386,636$
$\sum xy = 14847.9$
$n = 8$
$r = 0.602$
$H_0: \rho = 0$
$H_1: \rho \neq 0$
C. V. $= \pm 0.707 \qquad$ d. f. $= 6$
Decision: Do not reject. There is no significant linear relationship between fat and cholesterol. No regression should be done.

**22.**
$S_{est} = \sqrt{\frac{91 - (-1.918)(21) - 0.551(229)}{6-2}}$
$S_{est} = 1.129*$

**23.**
(For calculation purposes only, since no regression should be done.)

$S_{est} = \sqrt{\frac{386,636 - 110.12(1740) - 12.784(14,847.9)}{8-2}}$
$S_{est} = 29.47*$

**24.**
$y' = -1.918 + 0.551(7) = 1.939$ or 2

**24. continued**
$2 - 2.132(1.129)\sqrt{1 + \frac{1}{6} + \frac{6(11-9.833)^2}{6(621)-59^2}} < y$

$< 2 + 2.132(1.129)\sqrt{1 + \frac{1}{6} + \frac{6(11-9.833)^2}{6(621)-59^2}}$

$2 - 2.132(1.129)(1.095) < y < 2 + 2.132(1.129)(1.095)$

$-0.6 < y < 4.6$ or $0 < y < 5*$

**25.**
Since no regression should be done, the average of the $y'$ values is used: 217.5

**26.**
$y' = 98.7 + 3.82(3) + 6.51(1.5) = 119.9*$

**27.**
$R = \sqrt{\frac{(0.561)^2 + (0.714)^2 - 2(0.561)(0.714)(0.625)}{1 - (0.625)^2}}$

$R = 0.729*$

**28.**
$R_{adj}^2 = 1 - \left[ \frac{(1-0.774^2)(8-1)}{(8-2-1)} \right]$

$R_{adj}^2 = 0.439*$

*These answers may vary due to method of calculation and/or rounding.

Note: Graphs are not to scale and are intended to convey a general idea.

Answers may vary due to rounding, TI-83's, or computer programs.

EXERCISE SET 11-1

1.
The variance test compares a sample variance to a hypothesized population variance, while the goodness of fit test compares a distribution obtained from a sample with a hypothesized distribution.

3.
The expected values are computed based on what the null hypothesis states about the distribution.

5.
$H_0$: 82% of home-schooled students receive their education entirely at home, 12% attend school up to 9 hours per week, and 6% spend from 9 to 25 hours per week at school.
$H_1$: The proportions differ from those stated in the null hypothesis. (claim)

C. V. = 5.991    d. f. = 2    $\alpha = 0.05$

| O | E |
|---|---|
| 50 | 0.82(85) = 69.7 |
| 25 | 0.12(85) = 10.2 |
| 10 | 0.06(85) = 5.1 |

$$\chi^2 = \sum \frac{(O-E)^2}{E} = \frac{(50-69.7)^2}{69.7} + \frac{(25-10.2)^2}{10.2} +$$

$$\frac{(10-5.1)^2}{5.1} = 31.750$$

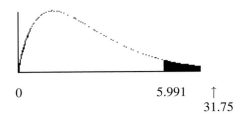

0                                5.991    ↑
                                         31.75

Reject the null hypothesis. There is enough evidence to support the claim that the proportions are different from those stated by the government.

7.
$H_0$: The proportions are distributed as follows: 30.6% purchased a small car, 45% purchased a mid-sized car, 7.3% purchased a large car, and 17.1% purchased a luxury car.
$H_1$: The distribution is not the same as stated in the null hypothesis. (claim)

C. V. = 7.815    d. f. = 3    $\alpha = 0.05$

| O | E |
|---|---|
| 25 | 0.306(100) = 30.6 |
| 50 | 0.45(100) = 45 |
| 10 | 0.073(100) = 7.3 |
| 15 | 0.171(100) = 17.1 |

$$\chi^2 = \sum \frac{(O-E)^2}{E} = \frac{(25-30.6)^2}{30.6} + \frac{(50-45)^2}{45}$$

$$+ \frac{(10-7.3)^2}{7.3} + \frac{(15-17.1)^2}{17.1} = 2.837$$

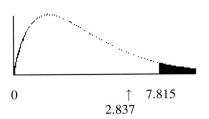

0                          ↑    7.815
                         2.837

Do not reject the null hypothesis. There is not enough evidence to support the claim that the proportions are different.

9.
$H_0$: The proportions are distributed as follows: safe - 35%, not safe - 52%, no opinion - 13%.
$H_1$: The distribution is not the same as stated in the null hypothesis. (claim)
C. V. = 9.210    d. f. = 2    $\alpha = 0.01$

| O | E |
|---|---|
| 40 | 0.35(120) = 42 |
| 60 | 0.52(120) = 62.4 |
| 20 | 0.13(120) = 15.6 |

$$\chi^2 = \frac{(40-42)^2}{42} + \frac{(60-62.4)^2}{62.4} + \frac{(20-15.6)^2}{15.6}$$

$$\chi^2 = 1.4286$$

9. continued

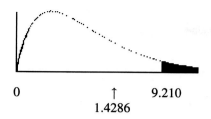

0        ↑     9.210
         1.4286

Do not reject the null hypothesis. There is not enough evidence to support the claim that the proportions are different.

11.
$H_0$: The distribution of loans is as follows: 21% - mortgages, 39% - autos, 20% - credit card, 12% - real estate, and 8% - miscellaneous. (claim)
$H_1$: The distribution is not the same as stated in the null hypothesis.

| O | E |
|---|---|
| 25 | 0.21(100) = 21 |
| 44 | 0.39(100) = 39 |
| 19 | 0.20(100) = 20 |
| 8 | 0.12(100) = 12 |
| 4 | 0.08(100) = 8 |

C. V. = 9.488    d. f. = 4    $\alpha = 0.05$

$$\chi^2 = \frac{(25-21)^2}{21} + \frac{(44-39)^2}{39} + \frac{(18-20)^2}{20} +$$

$$\frac{(8-12)^2}{12} + \frac{(4-8)^2}{8} = 4.9362$$

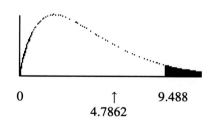

0       ↑     9.488
        4.7862

Do not reject the null hypothesis. There is not enough evidence to reject the claim that the distribution is the same.

13.
$H_0$: The method of payment for purchases is distributed as follows: 53% cash, 30% checks, 16% credit cards, and 1% no preference. (claim)

13. continued
$H_1$: The distribution is not the same as stated in the null hypothesis.

| O | E |
|---|---|
| 400 | 0.53(800) = 424 |
| 210 | 0.30(800) = 240 |
| 170 | 0.16(800) = 128 |
| 20 | 0.01(800) = 8 |

C. V. = 11.345    d. f. = 3    $\alpha = 0.01$

$$\chi^2 = \frac{(400-424)^2}{424} + \frac{(210-240)^2}{240} + \frac{(170-128)^2}{128}$$

$$+ \frac{(20-8)^2}{8} = 36.8897$$

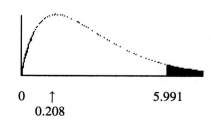

0              11.345   ↑
                   36.8897

Reject the null hypothesis. There is enough evidence to reject the claim that the distribution is the same as reported in the survey.

15.
$H_0$: The proportion of Internet users is the same for all groups.
$H_1$: The proportion of Internet users is not the same for all groups. (claim)

$$E = \frac{125}{3} = 41.67$$

C. V. = 5.991    d. f. = 2    $\alpha = 0.05$

$$\chi^2 = \sum \frac{(O-E)^2}{E} = \frac{(44-41.67)^2}{41.67} + \frac{(41-41.67)^2}{41.67}$$

$$+ \frac{(40-41.67)^2}{41.67}$$

$$\chi^2 = 0.208$$

0   ↑           5.991
    0.208

**15. continued**
Do not reject the null hypothesis. There is not enough evidence to support the claim that the proportions differ.

**17.**
$H_0$: The distribution of the ways people pay for their prescriptions is as follows: 60% used personal funds, 25% used insurance, and 15% used Medicare. (claim)
$H_1$: The distribution is not the same as stated in the null hypothesis.

$\alpha = 0.05$    d. f. $= 2$
P-value $> 0.05$
(TI: P-value $= 0.7164$)

| O | E |
|---|---|
| 32 | $0.6(50) = 30$ |
| 10 | $0.25(50) = 12.5$ |
| 8 | $0.15(50) = 7.5$ |

$\chi^2 = \sum \frac{(O-E)^2}{E} = \frac{(32-30)^2}{30} + \frac{(10-12.5)^2}{12.5} +$

$\frac{(8-7.5)^2}{7.5} = 0.667$

Do not reject the null hypothesis since P-value $> 0.05$. There is not enough evidence to reject the claim that the distribution is the same as stated in the null hypothesis. An implication of the results is that the majority of people are using their own money to pay for medications. A less expensive medication could help people financially.

**19.**
Answers will vary.

**EXERCISE SET 11-2**

**1.**
The independence test and the goodness of fit test both use the same formula for computing the test-value; however, the independence test uses a contingency table whereas the goodness of fit test does not.

**3.**
$H_0$: The variables are independent or not related.
$H_1$: The variables are dependent or related.

**5.**
The expected values are computed as (row total · column total) ÷ grand total.

**7.**
$H_0$: $p_1 = p_2 = p_3 = \cdots = p_n$
$H_1$: At least one proportion is different from the others.

**9.**
$H_0$: The number of endangered species is independent of the number of threatened species.
$H_1$: The number of endangered species is dependent upon the number of threatened species. (claim)

C. V. $= 9.488$    d. f. $= 4$    $\alpha = 0.05$

$E = \frac{\text{(row sum)(column sum)}}{\text{grand total}}$

$E_{1,1} = \frac{(247)(81)}{369} = 54.2195$

$E_{1,2} = \frac{(247)(91)}{369} = 60.9133$

$E_{1,3} = \frac{(247)(37)}{369} = 24.7669$

$E_{1,4} = \frac{(247)(23)}{369} = 15.3957$

$E_{1,5} = \frac{(247)(137)}{369} = 91.7046$

$E_{2,1} = \frac{(122)(81)}{369} = 26.7805$

$E_{2,2} = \frac{(122)(91)}{369} = 30.0867$

$E_{2,3} = \frac{(122)(37)}{369} = 12.2331$

$E_{2,4} = \frac{(122)(23)}{369} = 7.6043$

$E_{2,5} = \frac{(122)(137)}{369} = 45.2954$

| | Mammal | Bird |
|---|---|---|
| Endangered | 68(54.2195) | 76(60.9133) |
| Threatened | 13(26.7805) | 15(30.0867) |

| | Reptile | Amphibian |
|---|---|---|
| Endangered | 14(24.7669) | 13(15.3957) |
| Threatened | 23(12.2331) | 10(7.6043) |

| | Fish |
|---|---|
| Endangered | 76(91.7046) |
| Threatened | 61(45.2954) |

9. continued

$$\chi^2 = \sum \frac{(O-E)^2}{E} = \frac{(68-54.2195)^2}{54.2195} + \frac{(76-60.9133)^2}{60.9133}$$

$$+ \frac{(13-26.7805)^2}{26.7805} + \frac{(15-30.0867)^2}{30.0867} + \frac{(14-24.7669)^2}{24.7669}$$

$$+ \frac{(13-15.3957)^2}{15.3957} + \frac{(23-12.2331)^2}{12.2331} + \frac{(10-7.6043)^2}{7.6043}$$

$$+ \frac{(76-91.7046)^2}{91.7046} + \frac{(61-45.2954)^2}{45.2954}$$

$$\chi^2 = 45.3145$$

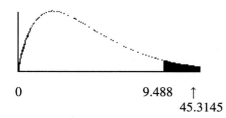

0        9.488    ↑
                 45.3145

Reject the null hypothesis. There is enough evidence to support the claim that there is a relationship between the type of vertebrate and whether it is endangered or threatened. The result is not different for $\alpha = 0.01$.

11.
$H_0$: The composition of the House of Representatives is independent of the state.
$H_1$: The composition of the House of Representatives is dependent upon the state. (claim)

C. V. = 7.815     d. f. = 3     $\alpha = 0.05$

$$E = \frac{\text{(row sum)(column sum)}}{\text{grand total}}$$

$$E_{1,1} = \frac{(203)(320)}{542} = 119.8524$$

$$E_{1,2} = \frac{(203)(222)}{542} = 83.1476$$

$$E_{2,1} = \frac{(98)(320)}{542} = 57.8598$$

$$E_{2,2} = \frac{(98)(222)}{542} = 40.1402$$

$$E_{3,1} = \frac{(100)(320)}{542} = 59.0406$$

$$E_{3,2} = \frac{(100)(222)}{542} 40.9594$$

$$E_{4,1} = \frac{(141)(320)}{542} = 83.2472$$

$$E_{4,2} = \frac{(141)(222)}{542} = 57.7528$$

11. continued

| State | Democrats | Republicans |
|-------|-----------|-------------|
| PA | 100(119.8524) | 103(83.1476) |
| OH | 39(57.8598) | 59(40.1402) |
| WV | 75(59.0406) | 25(40.9594) |
| MD | 106(83.2472) | 35(57.7528) |

$$\chi^2 = \sum \frac{(O-E)^2}{E} = \frac{(100-119.8524)^2}{119.8524} + \frac{(103-83.1476)^2}{83.1476}$$

$$+ \frac{(39-57.8598)^2}{57.8598} + \frac{(59-40.1402)^2}{40.1402} + \frac{(75-59.0406)^2}{59.0406}$$

$$+ \frac{(25-40.9594)^2}{40.9594} + \frac{(106-83.2472)^2}{83.2472} + \frac{(35-57.7528)^2}{57.7528}$$

$$\chi^2 = 48.7521$$

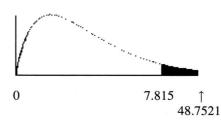

0             7.815    ↑
                       48.7521

Reject the null hypothesis. There is enough evidence to support the claim that the composition of the legislature is dependent upon the state.

13.
$H_0$: The occupation of state legislators is independent of the Congressional session.
$H_1$: The occupation of state legislators is dependent upon the Congressional session. (claim)

C. V. = 15.507     d. f. = 8     $\alpha = 0.05$

$$E_{1,1} = \frac{(16)(109)}{315} = 5.5365$$

$$E_{1,2} = \frac{(16)(104)}{315} = 5.2825$$

$$E_{1,3} = \frac{(16)(102)}{315} = 5.1810$$

$$E_{2,1} = \frac{(79)(109)}{315} = 27.3365$$

$$E_{2,2} = \frac{(79)(104)}{315} = 26.0825$$

$$E_{2,3} = \frac{(79)(102)}{315} = 25.581$$

$$E_{3,1} = \frac{(40)(109)}{315} = 13.8413$$

$$E_{3,2} = \frac{(40)(104)}{315} = 13.2063$$

13. continued

$E_{3,3} = \frac{(40)(102)}{315} = 12.9524$

$E_{4,1} = \frac{(170)(109)}{315} = 58.8254$

$E_{4,2} = \frac{(170)(104)}{315} = 56.1270$

$E_{4,3} = \frac{(170)(102)}{315} = 55.0476$

$E_{5,1} = \frac{(10)(109)}{315} = 3.4603$

$E_{5,2} = \frac{(10)(104)}{315} = 3.3016$

$E_{5,3} = \frac{(10)(102)}{315} = 3.2381$

|  | $109^{th}$ | $108^{th}$ | $107^{th}$ |
|---|---|---|---|
| Agriculture | 5(5.5365) | 5(5.2825) | 6(5.1810) |
| Business | 30(27.3365) | 25(26.0825) | 24(25.5810) |
| Education | 12(13.8413) | 12(13.2063) | 16(12.9524) |

|  | $109^{th}$ | $108^{th}$ | $107^{th}$ |
|---|---|---|---|
| Law | 58(58.8254) | 59(56.1270) | 53(55.0476) |
| Medicine | 4(3.4603) | 3(3.3016) | 3(3.2381) |

$\chi^2 = \sum \frac{(O-E)^2}{E} = \frac{(5-5.5365)^2}{5.5365} + \frac{(5-5.2825)^2}{5.2825}$

$+ \frac{(6-5.1810)^2}{5.1810} + \frac{(30-27.3365)^2}{27.3365} + \frac{(25-26.0825)^2}{26.0825}$

$+ \frac{(24-25.5810)^2}{25.5810} + \frac{(12-13.8413)^2}{13.8413} + \frac{(12-13.2063)^2}{13.2063}$

$+ \frac{(16-12.9524)^2}{12.9524} + \frac{(58-58.8254)^2}{58.8254} + \frac{(59-56.1270)^2}{56.1270}$

$+ \frac{(53-55.0476)^2}{55.0476} + \frac{(4-3.4603)^2}{3.4603} + \frac{(3-3.3016)^2}{3.3016}$

$+ \frac{(3-3.2381)^2}{3.2381} = 2.035$

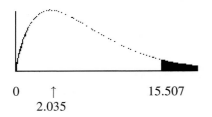

0    ↑      15.507
    2.035

Do not reject the null hypothesis. There is not enough evidence to support the claim that the occupation of U. S. senators is dependent upon the session.

15.
$H_0$: The program of study of a student is independent of the type of institution.

15. continued
$H_1$: The program of study of a student is dependent upon the type of institution. (claim)

C. V. $= 7.815$    d. f. $= 3$    $\alpha = 0.05$

$E_{1,1} = \frac{(88)(302)}{707} = 37.5898$

$E_{1,2} = \frac{(88)(405)}{707} = 50.4102$

$E_{2,1} = \frac{(441)(302)}{707} = 188.3762$

$E_{2,2} = \frac{(441)(405)}{707} = 252.6238$

$E_{3,1} = \frac{(87)(302)}{707} = 37.1627$

$E_{3,2} = \frac{(87)(405)}{707} = 49.8373$

$E_{4,1} = \frac{(91)(302)}{707} = 38.8713$

$E_{4,2} = \frac{(91)(405)}{707} = 52.1287$

|  | Two-year | Four-year |
|---|---|---|
| Agriculture | 36(37.5898) | 52(50.4102) |
| Criminal Justice | 210(188.3762) | 231(252.6238) |
| Lang/Lit | 28(37.1627) | 59(49.8373) |
| Math/Stat | 28(38.8713) | 63(52.1287) |

$\chi^2 = \sum \frac{(O-E)^2}{E} = \frac{(36-37.5898)^2}{37.5898} + \frac{(52-50.4102)^2}{50.4102}$

$+ \frac{(210-188.3762)^2}{188.3762} + \frac{(231-252.6238)^2}{252.6238} + \frac{(28-37.1627)^2}{37.1627}$

$+ \frac{(59-49.8373)^2}{49.8373} + \frac{(28-38.8713)^2}{38.8713} + \frac{(63-52.1287)^2}{52.1287}$

$\chi^2 = 13.702$

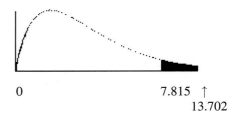

0            7.815 ↑
                 13.702

Reject the null hypothesis. There is enough evidence to conclude that the type of program is dependent on the type of institution.

**17.**

$H_0$: The type of video rented by a person is independent of the person's age.

$H_1$: The type of video a person rents is dependent on the person's age. (claim)

C. V. = 13.362   d. f. = 8   $\alpha = 0.10$

| Age | Doc. | Comedy | Mystery |
|------|---------|-----------|-----------|
| 12-20 | 14(6.588) | 9(13.433) | 8(10.979) |
| 21-29 | 15(8.075) | 14(16.467) | 9(13.458) |
| 30-38 | 9(14.663) | 21(29.9) | 39(24.438) |
| 39-47 | 7(9.775) | 22(19.933) | 17(16.292) |
| 48 + | 6(11.9) | 38(24.267) | 12(19.833) |

$$\chi^2 = \frac{(14-6.588)^2}{6.588} + \frac{(9-13.433)^2}{13.433} + \frac{(8-10.979)^2}{10.979}$$

$$+ \frac{(15-8.075)^2}{8.075} + \frac{(14-16.467)^2}{16.467} + \frac{(9-13.458)^2}{13.458}$$

$$+ \frac{(9-14.663)^2}{14.663} + \frac{(21-29.9)^2}{29.9} + \frac{(39-24.438)^2}{24.438}$$

$$+ \frac{(7-9.775)^2}{9.775} + \frac{(22-19.933)^2}{19.933} + \frac{(17-16.292)^2}{16.292}$$

$$+ \frac{(6-11.9)^2}{11.9} + \frac{(38-24.267)^2}{24.267} + \frac{(12-19.833)^2}{19.833}$$

$$\chi^2 = 46.733$$

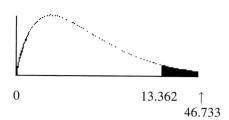

0                    13.362   ↑
                              46.733

Reject the null hypothesis. There is enough evidence to support the claim that the type of movie selected is related to the age of the customer.

**19.**

$H_0$: The type of snack purchased is independent of the gender of the consumer. (claim)

$H_1$: The type of snack purchased is dependent upon the gender of the consumer.

C. V. = 4.605   d. f. = 2

| Gender | Hot Dog | Peanuts | Popcorn |
|--------|-----------|-----------|-----------|
| Male | 12(13.265) | 21(15.388) | 19(23.347) |
| Female | 13(11.735) | 8(13.612) | 25(20.653) |

**19. continued**

$$\chi^2 = \sum \frac{(O-E)^2}{E} = \frac{(12-13.265)^2}{13.265} + \frac{(21-15.388)^2}{15.388}$$

$$+ \frac{(19-23.347)^2}{23.347} + \frac{(13-11.735)^2}{11.735} + \frac{(8-13.612)^2}{13.612}$$

$$+ \frac{(25-20.653)^2}{20.653} = 6.342$$

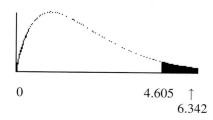

0                    4.605   ↑
                             6.342

Reject the null hypothesis. There is enough evidence to reject the claim that the type of snack chosen is independent of the gender of the individual.

**21.**

$H_0$: The type of book purchased by the individual is independent of the gender of the individual. (claim)

$H_1$: The type of book purchased by the individual is dependent on the gender of the individual.

$\alpha = 0.05$   d. f. = 2

| Gender | Mystery | Romance | Self-help |
|--------|------------|------------|------------|
| Male | 243(214.121) | 201(198.260) | 191(222.618) |
| Female | 135(163.879) | 149(151.740) | 202(170.382) |

$$\chi^2 = \sum \frac{(O-E)^2}{E} = \frac{(243-214.121)^2}{214.121} + \frac{(201-198.260)^2}{198.260}$$

$$+ \frac{(191-222.618)^2}{222.618} + \frac{(135-163.879)^2}{163.879}$$

$$+ \frac{(149-151.740)^2}{151.740} + \frac{(202-170.382)^2}{170.382} = 19.429$$

P-value < 0.005  (0.00006)
(TI: P-value = 0.00006)
Reject the null hypothesis since P-value < 0.05. There is enough evidence to reject the claim that the type of book purchased is independent of gender.

**23.**

$H_0$: $p_1 = p_2 = p_3 = p_4$  (claim)

$H_1$: At least one proportion is different.

C. V. = 7.815   d. f. = 3

**23. continued**

$E(\text{passed}) = \frac{120(167)}{120} = 41.75$

$E(\text{failed}) = \frac{120(313)}{120} = 78.25$

|        | Southside | West End | East Hills | Jefferson |
|--------|-----------|----------|------------|-----------|
| Passed | 49(41.75) | 38(41.75) | 46(41.75) | 34(41.75) |
| Failed | 71(78.25) | 82(78.25) | 74(78.25) | 86(78.25) |

$\chi^2 = \frac{(49-41.75)^2}{41.75} + \frac{(38-41.75)^2}{41.75} + \frac{(46-41.75)^2}{41.75}$

$+ \frac{(34-41.75)^2}{41.75} + \frac{(71-78.25)^2}{78.25} + \frac{(82-78.25)^2}{78.25}$

$+ \frac{(74-78.25)^2}{78.25} + \frac{(86-78.25)^2}{78.25} = 5.317$

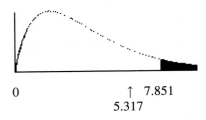

0          ↑ 7.851
           5.317

Do not reject the null hypothesis. There is not enough evidence to reject the claim that the proportions are equal.

**25.**

$H_0$: $p_1 = p_2 = p_3 = p_4$    (claim)
$H_1$: At least one proportion is different.

C. V. = 7.815    d. f. = 3

|            | Services | Manufacturing | Government | Other |
|------------|----------|---------------|------------|-------|
| 10 years ago | 33(30.6) | 13(15)     | 11(11.4)   | 3(3)  |
| Now        | 18(20.4) | 12(10)        | 8(7.6)     | 2(2)  |
| Total      | 51       | 25            | 19         | 5     |

$\chi^2 = \frac{(33-30.6)^2}{30.6} + \frac{(13-15)^2}{15} + \frac{(11-11.4)^2}{11.4} +$

$\frac{(3-3)^2}{3} + \frac{(18-20.4)^2}{20.4} + \frac{(12-10)^2}{10} + \frac{(8-7.6)^2}{7.6}$

$+ \frac{(2-2)^2}{2} = 1.172$

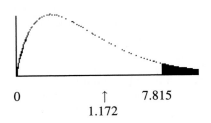

0          ↑     7.815
           1.172

**25. continued**

Do not reject the null hypothesis. There is not enough evidence to reject the claim that the proportions are the same. Since the survey was done in Pennsylvania, it is doubtful that it can be generalized to the population of the United States.

**27.**

$H_0$: $p_1 = p_2 = p_3 = p_4 = p_5$
$H_1$: At least one proportion is different.

C. V. = 9.488    d. f. = 4

$E(\text{yes}) = \frac{(104)(75)}{375} = 20.8$

$E(\text{no}) = \frac{(271)(75)}{375} = 54.2$

|     | 18        | 19        | 20        |
|-----|-----------|-----------|-----------|
| Yes | 19(20.8)  | 18(20.8)  | 23(20.8)  |
| No  | 56(54.2)  | 57(54.2)  | 52(54.2)  |

|     | 21        | 22        |
|-----|-----------|-----------|
| Yes | 31(20.8)  | 13(20.8)  |
| No  | 44(54.2)  | 62(54.2)  |

$\chi^2 = \frac{(19-20.8)^2}{20.8} + \frac{(18-20.8)^2}{20.8} + \frac{(23-20.8)^2}{20.8} +$

$\frac{(31-20.8)^2}{20.8} + \frac{(13-20.8)^2}{20.8} + \frac{(56-54.2)^2}{54.2} + \frac{(57-54.2)^2}{54.2}$

$+ \frac{(52-54.2)^2}{54.2} + \frac{(44-54.2)^2}{54.2} + \frac{(62-54.2)^2}{54.2}$

$\chi^2 = 12.028$

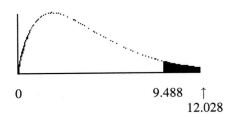

0          9.488   ↑
             12.028

Reject the null hypothesis. There is enough evidence to conclude that the proportions are different.

**29.**

$H_0$: $p_1 = p_2 = p_3 = p_4$    (claim)
$H_1$: At least one proportion is different.

$\alpha = 0.05$    d. f. = 3

$E(\text{on bars}) = \frac{30(62)}{120} = 15.5$

**29. continued**

$E(\text{not on bars}) = \frac{30(58)}{120} = 14.5$

|     | N | S | E | W |
|-----|-----|-----|-----|-----|
| on | 15(15.5) | 18(15.5) | 13(15.5) | 16(15.5) |
| off | 15(14.5) | 12(14.5) | 17(14.5) | 14(14.5) |

$\chi^2 = \frac{(15-15.5)^2}{15.5} + \frac{(18-15.5)^2}{15.5} + \frac{(13-15.5)^2}{15.5} +$

$\frac{(16-15.5)^2}{15.5} + \frac{(15-14.5)^2}{14.5} + \frac{(12-14.5)^2}{14.5} +$

$\frac{(17-14.5)^2}{14.5} + \frac{(14-14.5)^2}{14.5} = 1.734$

P-value > 0.10  (0.629)
(TI: P-value = 0.6291)

Do not reject the null hypothesis since P-value > 0.05. There is not enough evidence to reject the claim that the proportions are the same.

**31.**
$H_0$: $p_1 = p_2 = p_3$ (claim)
$H_1$: At least one proportion is different.

C. V. = 4.605    d. f. = 2

$E(\text{list}) = \frac{96(219)}{288} = 73$

$E(\text{no list}) = \frac{96(69)}{288} = 23$

|     | A | B | C |
|-----|-----|-----|-----|
| list | 77(73) | 74(73) | 68(73) |
| no list | 19(23) | 22(23) | 28(23) |

$\chi^2 = \frac{(77-73)^2}{73} + \frac{(74-73)^2}{73} + \frac{(68-73)^2}{73}$

$+ \frac{(19-23)^2}{23} + \frac{(22-23)^2}{23} + \frac{(28-23)^2}{23}$

$\chi^2 = 2.401$

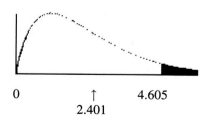

0                    ↑              4.605
                  2.401

Do not reject the null hypothesis. There is not enough evidence to reject the claim that the proportions are the same.

**33.**

$\chi^2 = \frac{(|O-E|-0.5)^2}{E} = \frac{(|12-9.6|-0.5)^2}{9.6}$

$+ \frac{(|15-17.4|-0.5)^2}{17.4} + \frac{(|9-11.4|-0.5)^2}{11.4}$

$+ \frac{(|23-20.6|-0.5)^2}{20.6}$

$= \frac{3.61}{9.6} + \frac{3.61}{17.4} + \frac{3.61}{11.4} + \frac{3.61}{20.6}$

$= 0.376 + 0.207 + 0.317 + 0.175 = 1.075$

**REVIEW EXERCISES - CHAPTER 11**

**1.**
$H_0$: People show no preference for the day of the week that they do their shopping. (claim)
$H_1$: People show a preference for the day of the week that they do their shopping.

C. V. = 12.592    d. f. = 6
$E = \frac{400}{7} = 57.14$

$\chi^2 = \sum \frac{(O-E)^2}{E} = \frac{(28-57.14)^2}{57.14} + \frac{(16-57.14)^2}{57.14}$

$+ \frac{(20-57.14)^2}{57.14} + \frac{(26-57.14)^2}{57.14} + \frac{(74-57.14)^2}{57.14}$

$+ \frac{(96-57.14)^2}{57.14} + \frac{(140-57.14)^2}{57.14} = 237.15$

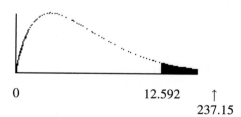

0                         12.592        ↑
                                     237.15

Reject the null hypothesis. There is enough evidence to reject the claim that shoppers have no preference for the day of the week that they do their shopping. Retail merchants could plan for more shoppers on Fridays and Saturdays than they will have on other days of the week.

**3.**
$H_0$: Opinion is independent of gender.
$H_1$: Opinion is dependent on gender. (claim)
C. V. = 4.605    d. f. = 2

3. continued

| Gender | Yes | No | Undecided |
|--------|-----|-----|-----------|
| Men | 114(120.968) | 30(22.258) | 6(6.774) |
| Women | 136(129.032) | 16(23.742) | 8(7.226) |

$$\chi^2 = \frac{(114-120.968)^2}{120.968} + \frac{(30-22.258)^2}{22.258} + \frac{(6-6.774)^2}{6.774}$$

$$+ \frac{(136-129.032)^2}{129.032} + \frac{(16-23.742)^2}{23.742} + \frac{(8-7.226)^2}{7.226}$$

$$\chi^2 = 6.16$$

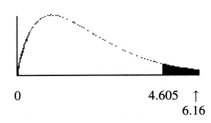

0          4.605  ↑
                  6.16

Reject the null hypothesis. There is enough evidence to support the claim that opinion is dependent on gender.

5.

$H_0$: The type of investment is independent of the age of the investor.
$H_1$: The type of investment is dependent upon the age of the investor. (claim)

C. V. = 9.488     d. f. = 4

| Age | Large | Small | Inter. |
|-----|-------|-------|--------|
| 45 | 20(28.18) | 10(15.45) | 10(15.45) |
| 65 | 42(33.82) | 24(18.55) | 24(18.55) |

| Age | CD | Bond |
|-----|-----|------|
| 45 | 15(9.55) | 45(31.36) |
| 65 | 6(11.45) | 24(37.64) |

$$\chi^2 = \frac{(20-28.18)^2}{28.18} + \frac{(10-15.45)^2}{15.45} + \frac{(10-15.45)^2}{15.45}$$

$$+ \frac{(15-9.55)^2}{9.55} + \frac{(45-31.36)^2}{31.36} + \frac{(42-33.82)^2}{33.82} +$$

$$\frac{(24-18.55)^2}{18.55} + \frac{(24-18.55)^2}{18.55} + \frac{(6-11.45)^2}{11.45} +$$

$$\frac{(24-37.64)^2}{37.64} = 27.998 \text{ or } 28.0$$

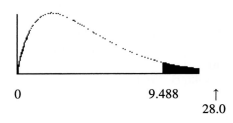

0          9.488    ↑
                    28.0

5. continued

Reject the null hypothesis. There is enough evidence to support the claim that the type of investment is dependent on age.

7.

$H_0$: $p_1 = p_2 = p_3$  (claim)
$H_1$: At least one proportion is different.

$\alpha = 0.01$     d. f. = 2

$$E(\text{work}) = \frac{80(114)}{240} = 38$$

$$E(\text{don't work}) = \frac{80(126)}{240} = 42$$

| | 16 | 17 | 18 |
|-----------|----|----|----|
| work | 45(38) | 31(38) | 38(38) |
| don't work | 35(42) | 49(42) | 42(42) |

$$\chi^2 = \frac{(45-38)^2}{38} + \frac{(31-38)^2}{38} + \frac{(38-38)^2}{38}$$

$$+ \frac{(35-42)^2}{42} + \frac{(49-42)^2}{42} + \frac{(42-42)^2}{42} = 4.912$$

$0.05 < $ P-value $< 0.10$  (0.086)

Do not reject the null hypothesis since P-value > 0.01. There is not enough evidence to reject the claim that the proportions are the same.

9.

$H_0$: $p_1 = p_2 = p_3 = p_4 = p_5$
$H_1$: At least one proportion is different.

$$E(\text{insurance}) = \frac{832(200)}{1000} = 166.4$$

$$E(\text{no insurance}) = \frac{168(200)}{1000} = 33.6$$

| | Va. | Ill. | D.C. |
|-----|-----|------|------|
| Yes | 171(166.4) | 171(166.4) | 170(166.4) |
| No | 29(33.6) | 29(33.6) | 30(33.6) |

| | Wyo. | Tex. |
|-----|------|------|
| Yes | 168(166.4) | 152(166.4) |
| No | 32(33.6) | 48(33.6) |

$$\chi^2 = \frac{(171-166.4)^2}{166.4} + \frac{(171-166.4)^2}{166.4}$$

$$+ \frac{(170-166.4)^2}{166.4} + \frac{(168-166.4)^2}{166.4} + \frac{(152-166.4)^2}{166.4}$$

$$+ \frac{(29-33.6)^2}{33.6} + \frac{(29-33.6)^2}{33.6} + \frac{(30-33.6)^2}{33.6}$$

$$+ \frac{(32-33.6)^2}{33.6} + \frac{(48-33.6)^2}{33.6}$$

9. continued
$\chi^2 = 9.487$

P-value $= 0.06$
(TI: P-value $= 0.050$)

Reject the null hypothesis since P-value
$< 0.10$. There is enough evidence to
conclude that the proportions are not all the
same.

## CHAPTER 11 QUIZ

1. False, it is one-tailed right.
2. True
3. False, there is little agreement between observed and expected frequencies.
4. c.
5. b.
6. d.
7. 6
8. independent
9. right
10. at least five

11. $H_0$: The reasons why people lost their jobs are equally distributed.   (claim)
$H_1$: The reasons why people lost their jobs are not equally distributed.
C. V. $= 5.991$   d. f. $= 2$   E $= 24$
$\chi^2 = \sum \frac{(O-E)^2}{E} = 2.33$
Do not reject the null hypothesis. There is not enough evidence to reject the claim that the reasons why people lost their jobs are equally distributed. The results could have been different 10 years ago since different factors of the economy existed then.

12. $H_0$: Takeout food is consumed according to the following distribution: at home - 53%, in the car - 19%, at work - 14%, other - 14%. (claim)
$H_1$: The distribution is different from that stated in the null hypothesis.
C. V. $= 11.345$   d. f. $= 3$
$\chi^2 = \sum \frac{(O-E)^2}{E} = 5.271$
Do not reject the null hypothesis. There is not enough evidence to reject the claim that the distribution is as stated. Fast-food restaurants may want to make their advertisements appeal to those who like to take their food home to eat.

13. $H_0$: College students show the same preference for shopping channels as those surveyed.
$H_1$: College students show a different preference for shopping channels. (claim)
C. V. $= 7.815$   d. f. $= 3$
$\chi^2 = 21.789$
Reject the null hypothesis. There is enough evidence to support the claim that college students show a different preference for shopping channels.

14. $H_0$: The number of commuters is distributed as follows: alone - 75.7%, carpooling - 12.2%, public transportation - 4.7%, walking - 2.9%, other - 1.2%, and working at home - 3.3%.
$H_1$: The proportions are different from the null hypothesis. (claim)
C. V. $= 11.071$   d. f. $= 5$
$\chi^2 = 41.269$
Reject the null hypothesis. There is enough evidence to support the claim that the distribution is different from the one stated in the null hypothesis.

15. $H_0$: Ice cream flavor is independent of the gender of the purchaser. (claim)
$H_1$: Ice cream flavor is dependent upon the gender of the purchaser.
C. V. $= 7.815$   d. f. $= 3$
$\chi^2 = 7.198$
Do not reject the null hypothesis. There is not enough evidence to reject the claim that ice cream flavor is independent of the gender of the purchaser.

16. $H_0$: The type of pizza ordered is independent of the age of the purchaser.
$H_1$: The type of pizza ordered is dependent on the age of the purchaser. (claim)
$\alpha = 0.10$   d. f. $= 9$
$\chi^2 = 107.3$
P-value $< 0.005$
Reject the null hypothesis since P-value $< 0.10$. There is enough evidence to support the claim that the type of pizza is related to the age of the purchaser.

17. $H_0$: The color of the pennant purchased is independent of the gender of the purchaser. (claim)
$H_1$: The color of the pennant purchased is dependent on the gender of the purchaser.
C. V. $= 4.605$   d. f. $= 2$

17. continued
$\chi^2 = 5.6$
Reject the null hypothesis. There is enough
evidence to reject the claim that the color of
the pennant purchased is independent of the
gender of the purchaser.

18. $H_0$: The opinion of the children on the
use of the tax credit is independent of the
gender of the children.
$H_1$: The opinion of the children on the use
of the tax credit is dependent upon the
gender of the children. (claim)
C. V. $= 4.605$    d. f. $= 2$
$\chi^2 = 1.534$
Do not reject the null hypothesis. There is
not enough evidence to support the claim
that the opinion of the children is dependent
upon their gender.

19. $H_0$: $p_1 = p_2 = p_3$ (claim)
$H_1$: At least one proportion is different from
the others.
C. V. $= 4.605$    d. f. $= 2$
$\chi^2 = 6.711$
Reject the null hypothesis. There is enough
evidence to reject the claim that the
proportions are equal. It seems that more
women are undecided about their jobs.
Perhaps they want better income or greater
chances of advancement.

Note: Graphs are not to scale and are intended to convey a general idea. Answers may vary due to rounding.

EXERCISE SET 12-1

**1.**

The analysis of variance using the F-test can be used to compare 3 or more means.

**3.**

The populations from which the samples were obtained must be normally distributed. The samples must be independent of each other. The variances of the populations must be equal.

**5.**

$$F = \frac{s_B^2}{s_W^2}$$

**7.**

Scheffe′ Test and the Tukey Test

**9.**

$H_0$: $\mu_1 = \mu_2 = \mu_3$

$H_1$: At least one mean is different from the others. (claim)

C. V. = 4.26     $\alpha = 0.05$

d. f. N. = 2     d. f. D. = 9

$\overline{X}_1 = 1.888$     $s_1 = 0.535$

$\overline{X}_2 = 2.224$     $s_2 = 0.1328$

$\overline{X}_3 = 3.525$     $s_3 = 0.1344$

$$\overline{X}_{GM} = \frac{27.61}{12} = 2.301$$

$$s_B^2 = \frac{5(1.888-2.301)^2+5(2.224-2.301)^2+2(3.525-2.301)^2}{3-1}$$

$$s_B^2 = 1.9394$$

$$s_W^2 = \frac{4(0.535)^2+4(0.1328)^2+1(0.1344)^2}{4+4+1}$$

$$s_W^2 = 0.1371$$

$F = \frac{1.9394}{0.1371} = 14.146$

(TI: F = 14.1489)

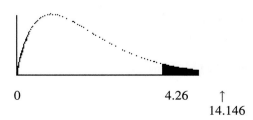

0                    4.26       ↑

                              14.146

**9. continued**

Reject the null hypothesis. There is enough evidence to conclude that at least one mean is different from the others.

**11.**

$H_0$: $\mu_1 = \mu_2 = \mu_3$

$H_1$: At least one mean is different from the others. (claim)

C. V. = 3.98     $\alpha = 0.05$

d. f. N = 2     d. f. D = 11

$$\overline{X}_{GM} = \frac{52414}{14} = 3743.857$$

$$s_B^2 = 3,633,540.88$$

$$s_W^2 = 1,330,350$$

$F = \frac{3633540.88}{1330350} = 2.7313$

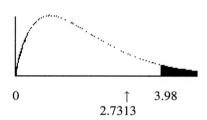

0                    ↑      3.98

              2.7313

Do not reject the null hypothesis. There is not enough evidence to support the claim that at least one mean is different from the others.

**13.**

$H_0$: $\mu_1 = \mu_2 = \mu_3$

$H_1$: At least one mean is different. (claim)

$k = 3$     $N = 18$     d.f.N. $= 2$     d.f.D. $= 15$

CV $= 3.68$

$\overline{X}_1 = 7$          $s_1^2 = 1.37$

$\overline{X}_2 = 8.12$     $s_2^2 = 0.64$

$\overline{X}_3 = 5.23$     $s_3^2 = 2.66$

$$\overline{X}_{GM} = 6.7833$$

$$s_B^2 = \frac{6(7-6.78)^2}{2} + \frac{6(8.12-6.78)^2}{2}$$

$$+ \frac{6(5.23-6.78)^2}{2} = 12.7$$

$$s_W^2 = \frac{5(1.37)+5(0.64)+5(2.66)}{5+5+5}$$

$$s_W^2 = \frac{23.35}{15} = 1.56$$

13. continued

$$F = \frac{12.7}{1.56} = 8.14$$

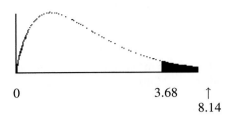

0          3.68     ↑
                8.14

Reject the null hypothesis. There is enough evidence to support the claim that at least one mean is different.

15.

$H_0$: $\mu_1 = \mu_2 = \mu_3$ (claim)
$H_1$: At least one mean is different from the others.

C. V. = 4.10          $\alpha = 0.10$
d. f. N = 2          d. f. D = 10

$\overline{X}_1 = 35.4$          $s_1^2 = 351.8$
$\overline{X}_2 = 68.75$          $s_2^2 = 338.25$
$\overline{X}_3 = 44.25$          $s_3^2 = 277.583$

$\overline{X}_{GM} = \frac{629}{13} = 48.385$

$s_B^2 = \frac{\sum n_i(\overline{X}_i - \overline{X}_{GM})^2}{k-1}$

$s_B^2 = \frac{5(35.4-48.385)^2}{2} + \frac{4(68.75-48.385)^2}{2}$

$+ \frac{4(44.25-48.385)^2}{2} = 1285.188$

$s_W^2 = \frac{\sum(n_i-1)s_i^2}{\sum(n_i-1)}$

$s_W^2 = \frac{4(351.8) + 3(338.25) + 3(277.583)}{4+3+3}$

$= 325.47$

$F = \frac{s_B^2}{s_W^2} = \frac{1285.188}{325.47} = 3.9487$

Do not reject the null hypothesis. There is enough evidence to reject the claim that the means are the same.

17.

$H_0$: $\mu_1 = \mu_2 = \mu_3$
$H_1$: At least one mean is different from the others.  (claim)

C. V. = 2.61          $\alpha = 0.10$

17. continued

d. f. N = 2          d. f. D = 19

$\overline{X}_1 = 233.33$          $s_1 = 28.225$
$\overline{X}_2 = 203.125$          $s_2 = 39.364$
$\overline{X}_3 = 155.625$          $s_3 = 28.213$

$\overline{X}_{GM} = 194.091$

$s_B^2 = \frac{21{,}729.735}{2} = 10{,}864.8675$

$s_W^2 = \frac{20{,}402.083}{19} = 1073.794$

$F = \frac{s_B^2}{s_W^2} = \frac{10{,}864.8675}{1073.794} = 10.12$

P-value = 0.00102
Reject since P-value < 0.10.

19.

$H_0$: $\mu_1 = \mu_2 = \mu_3 = \mu_4$
$H_1$: At least one mean is different.  (claim)

C. V. = 3.24          $\alpha = 0.05$
d. f. N = 3          d. f. D = 16

$\overline{X}_{GM} = \frac{308{,}325}{20} = 15{,}416.25$

$\overline{X}_1 = 14{,}394.6$          $s_1 = 1253.424$
$\overline{X}_2 = 14{,}668.2$          $s_2 = 2367.532$
$\overline{X}_3 = 14{,}275.2$          $s_3 = 821.006$
$\overline{X}_4 = 18{,}327$          $s_4 = 2415.376$

$s_B^2 = \frac{56{,}889{,}041}{3} = 18{,}963{,}013.7$

$s_W^2 = \frac{54{,}737{,}476.8}{16} = 3{,}421{,}092.3$

$F = \frac{18{,}963{,}013.7}{3{,}421{,}092.3} = 5.543$

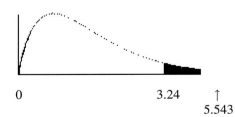

0          3.24     ↑
                5.543

Reject the null hypothesis. There is not enough evidence to support the claim that at least one mean is different. Students may have had discipline problems.

EXERCISE SET 12-2

**1.**
The Scheffe' and Tukey tests are used.

**3.**
Scheffe' Test
C. V. = 8.52

$$F_s = \frac{(\overline{X}_i - \overline{X}_j)^2}{s_W^2(\frac{1}{n_i} + \frac{1}{n_j})}$$

For $\overline{X}_1$ vs $\overline{X}_2$

$$F_S = \frac{(1.888 - 2.224)^2}{0.13707\left(\frac{1}{5} + \frac{1}{5}\right)} = \frac{0.112896}{0.054828} = 2.059$$

For $\overline{X}_1$ vs $\overline{X}_3$

$$F_S = \frac{(1.888 - 3.525)^2}{0.13707\left(\frac{1}{5} + \frac{1}{2}\right)} = \frac{2.679769}{0.095949} = 27.929$$

For $\overline{X}_2$ vs $\overline{X}_3$

$$F_S = \frac{(2.224 - 3.525)^2}{0.13707\left(\frac{1}{5} + \frac{1}{2}\right)} = \frac{1.692601}{0.0959504} = 17.640$$

There is a significant difference between $\overline{X}_1$ and $\overline{X}_3$ and between $\overline{X}_2$ and $\overline{X}_3$.

**5.**
Tukey Test:
C. V. = 3.29

$\overline{X}_1$ vs $\overline{X}_2$:

$$q = \frac{7 - 8.12}{\sqrt{\frac{1.56}{6}}} = -2.196$$

$\overline{X}_1$ vs $\overline{X}_3$:

$$q = \frac{7 - 5.23}{\sqrt{\frac{1.56}{6}}} = 3.47$$

$\overline{X}_2$ vs $\overline{X}_3$:

$$q = \frac{8.12 - 5.23}{\sqrt{\frac{1.56}{6}}} = 5.667$$

There is a significant difference between $\overline{X}_1$ and $\overline{X}_3$ and between $\overline{X}_2$ and $\overline{X}_3$.

**7.**
Scheffe' Test
C. V. = 5.22

**7.** continued
$\overline{X}_1$ vs $\overline{X}_2$:

$$F_s = \frac{(\overline{X}_i - \overline{X}_j)^2}{s_W^2(\frac{1}{n_i} + \frac{1}{n_j})} = \frac{(233.33 - 203.125)^2}{1073.776(\frac{1}{6} + \frac{1}{8})}$$

$$F_s = 2.91$$

$\overline{X}_1$ vs $\overline{X}_3$:

$$F_s = \frac{(233.33 - 155.625)^2}{1073.776(\frac{1}{6} + \frac{1}{8})} = 19.28$$

$\overline{X}_2$ vs $\overline{X}_3$ :

$$F_s = \frac{(203.125 - 155.625)^2}{1073.776(\frac{1}{8} + \frac{1}{8})} = 8.40$$

There is a significant difference between $\overline{X}_1$ and $\overline{X}_3$ and between $\overline{X}_2$ and $\overline{X}_3$.

**9.**
Tukey Test:
C. V. = 4.05

$$q = \frac{\overline{X}_i - \overline{X}_j}{\sqrt{\frac{s_W^2}{n}}}$$

$\overline{X}_1$ vs $\overline{X}_2$:

$$q = \frac{(14,394.6 - 14,668.2)}{\sqrt{\frac{3,421,092.3}{5}}} = -0.331$$

$\overline{X}_1$ vs $\overline{X}_3$:

$$q = \frac{(14,394.6 - 14,275.2)}{\sqrt{\frac{3,421,092.3}{5}}} = 0.144$$

$\overline{X}_1$ vs $\overline{X}_4$:

$$q = \frac{(14,394.6 - 18,327)}{\sqrt{\frac{3,421,092.3}{5}}} = -4.754$$

$\overline{X}_2$ vs $\overline{X}_3$ :

$$q = \frac{(14,668.2 - 14,275.2)}{\sqrt{\frac{3,421,092.3}{5}}} = 0.475$$

$\overline{X}_2$ vs $\overline{X}_4$:

$$q = \frac{(14,668.2 - 18,327)}{\sqrt{\frac{3,421,092.3}{5}}} = -4.42$$

$\overline{X}_3$ vs $\overline{X}_4$:

$$q = \frac{(14,275.2 - 18,327)}{\sqrt{\frac{3,421,092.3}{5}}} = -4.899$$

9. continued
There is a significant difference between $\overline{X}_1$ and $\overline{X}_4$, $\overline{X}_2$ and $\overline{X}_4$, and $\overline{X}_3$ and $\overline{X}_4$.

11.
$H_0$: $\mu_1 = \mu_2 = \mu_3$
$H_1$: At least one mean is different from the others. (claim)
C. V. = 3.47      $\alpha = 0.05$
d. f. N. = 2      d. f. D. = 21

$\overline{X}_{GM} = 4.554$   $s_B^2 = 9.82113$   $s_W^2 = 4.93225$

$F = \frac{9.82113}{4.93225} = 1.9912$

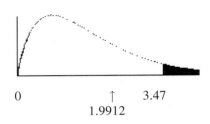

0              $\uparrow$      3.47
                1.9912

Do not reject the null hypothesis. There is not enough evidence to support the claim that at least one mean is different from the others.

13.
$H_0$: $\mu_1 = \mu_2 = \mu_3 = \mu_4$
$H_1$: At least one mean is different. (claim)

C. V. = 5.29      $\alpha = 0.01$
d. f. N = 3       d. f. D = 16

$\overline{X}_{GM} = \frac{42}{20} = 2.1$

$s_B^2 = \frac{10.2}{3} = 3.4$

$s_W^2 = \frac{85.6193}{16} = 5.35$

$F = \frac{3.4}{5.35} = 0.636$

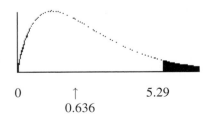

0       $\uparrow$          5.29
      0.636

Do not reject the null hypothesis. There is not enough evidence to support the claim

13. continued
that at least one mean is different. Students may have had discipline problems. Parents may not like the regular school district, etc.

EXERCISE SET 12-3

1.
The two-way ANOVA allows the researcher to test the effects of two independent variables and a possible interaction effect. The one-way ANOVA can test the effects of one independent variable only.

3.
The mean square values are computed by dividing the sum of squares by the corresponding degrees of freedom.

5.
a. d. f.$_A$ = $(3 - 1)$ = 2 for factor A
b. d. f.$_B$ = $(2 - 1)$ = 1 for factor B
c. d. f.$_{AxB}$ = $(3 - 1)(2 - 1)$ = 2
d. d. f.$_{within}$ = $3 \cdot 2(5 - 1)$ = 24

7.
The two types of interactions that can occur are ordinal and disordinal.

9.
a. The lines will be parallel or approximately parallel. They could also coincide.
b. The lines will not intersect and they will not be parallel.
c. The lines will intersect.

11.
$H_0$: There is no interaction effect between temperature and level of humidity.
$H_1$: There is an interaction effect between temperature and level of humidity.

$H_0$: There is no difference in mean length of effectiveness with respect to humidity.
$H_1$: There is a difference in mean length of effectiveness with respect to humidity.

$H_0$: There is no difference in mean length of effectiveness based on temperature.
$H_1$: There is a difference in mean length of effectiveness based on temperature.

11. continued

ANOVA SUMMARY TABLE

| Source | SS | d. f. | MS | F |
|--------|----|----|----|----|
| Humidity | 280.3333 | 1 | 280.3333 | 18.3825 |
| Temperature | 3 | 1 | 3 | 0.1967 |
| Interaction | 65.3333 | 1 | 65.3333 | 4.2842 |
| Within | 122 | 8 | 15.25 | |
| Total | 470.6667 | 11 | | |

The critical value at $\alpha = 0.05$ with d. f. N = 1 and d. f. D = 8 is 5.318 for $F_A$, $F_B$, and $F_{A \times B}$.

Since the only F test value that exceeds the critical value is the one for humidity, there is sufficient evidence to conclude that there is a difference in mean length of effectiveness based on the humidity level.

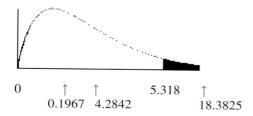

0          ↑      ↑          5.318     ↑
      0.1967   4.2842              18.3825

13.
$H_0$: There is no interaction effect between the type of flour and the sweetening agent in the glaze.
$H_1$: There is an interaction effect between the type of flour and the sweetening agent in the glaze.

$H_0$: There is no difference in mean sales based on the type of flour used.
$H_1$: There is a difference in mean sales based on the type of flour used.

$H_0$: There is no difference in mean sales based on the type of sweetener used in the glaze.
$H_1$: There is a difference in mean sales based on the type of sweetener used in the glaze.

ANOVA SUMMARY TABLE

| Source | SS | d. f. | MS | F |
|--------|----|----|----|----|
| Flour | 0.5625 | 1 | 0.5625 | 0.0062 |
| Sweetener | 1105.563 | 1 | 1105.563 | 12.2812 |
| Interaction | 0.5625 | 1 | 0.5625 | 0.0062 |
| Within | 1080.25 | 12 | 90.0208 | |
| Total | 2186.938 | 15 | | |

13. continued
The critical value at $\alpha = 0.05$ with d. f. N = 1 and d. f. D = 12 is 4.747. Since the F test value for the sweetener is greater than the critical value, it can be concluded that there is a difference in mean sales based on the type of sweetener used in the glaze.

15.
$H_0$: There is no interaction effect between the ages of the salespersons and the products they sell on the monthly sales.
$H_1$: There is an interaction effect between the ages of the salespersons and the products they sell on the monthly sales.

$H_0$: There is no difference in the means of the monthly sales of the two age groups.
$H_1$: There is a difference in the means of the monthly sales of the two age groups.

$H_0$: There is no difference among the means of the sales for the different products.
$H_1$: There is a difference among the means of the sales for the different products.

ANOVA SUMMARY TABLE

| Source | SS | d. f. | MS | F |
|--------|----|----|----|----|
| Age | 168.033 | 1 | 168.033 | 1.567 |
| Product | 1762.067 | 2 | 881.034 | 8.215 |
| Interaction | 7955.267 | 2 | 3977.634 | 37.087 |
| Error | 2574.000 | 24 | 107.250 | |
| Total | 12459.367 | 29 | | |

At $\alpha = 0.05$, the critical values are:

For age, d. f. N = 1, d. f. D = 24, C. V. = 4.26

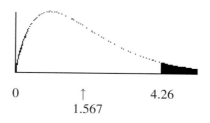

0              ↑              4.26
          1.567

For product and interaction, d. f. N = 2, d. f. D = 24, and C. V. = 3.40

15. continued

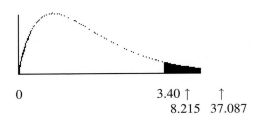

0           3.40 ↑    ↑
               8.215   37.087

The null hypotheses for the interaction effect and for the type of product sold are rejected since the F test values exceed the critical value, 3.40. The cell means are:

| Age | Pools | Spas | Saunas |
|---|---|---|---|
| over 30 | 38.8 | 28.6 | 55.4 |
| 30 & under | 21.2 | 68.6 | 18.8 |

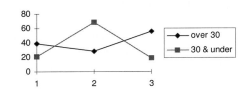

Since the lines cross, there is a disordinal interaction hence there is an interaction effect between the age of the sales person and the type of products sold.

## REVIEW EXERCISES - CHAPTER 12

1.
$H_0: \mu_1 = \mu_2 = \mu_3$ (claim)
$H_1:$ At least one mean is different from the others.
C. V. = 5.39    $\alpha = 0.01$
d. f. N = 2     d. f. D = 33

$\overline{X}_1 = 620.5$      $s_1^2 = 5445.91$
$\overline{X}_2 = 610.17$     $s_2^2 = 22{,}108.7$
$\overline{X}_3 = 477.83$     $s_3^2 = 5280.33$

$\overline{X}_{GM} = \frac{20{,}502}{36} = 569.5$

$s_B^2 = \frac{151{,}890.667}{2} = 75{,}945.333$

$s_W^2 = \frac{361{,}184.333}{33} = 10{,}944.9798$

$F = \frac{s_B^2}{s_W^2} = \frac{75{,}945.333}{10{,}944.9798} = 6.94$

1. continued

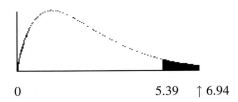

0           5.39    ↑ 6.94

Reject. At least one mean is different.

Tukey Test      C. V. = 4.45 using (3, 33)

$\overline{X}_1$ vs $\overline{X}_2$

$q = \dfrac{\overline{X}_1 - \overline{X}_2}{\sqrt{\frac{s_W^2}{n}}} = \dfrac{620.5 - 610.17}{\sqrt{\frac{10{,}944.98}{12}}} = 0.342$

$\overline{X}_1$ vs $\overline{X}_3$

$q = \dfrac{620.5 - 477.83}{\sqrt{\frac{10{,}944.98}{12}}} = 4.72$

$\overline{X}_2$ vs $\overline{X}_3$

$q = \dfrac{610.17 - 477.83}{\sqrt{\frac{10{,}944.98}{12}}} = 4.38$

There is a significant difference between $\overline{X}_1$ and $\overline{X}_3$.

3.
$H_0: \mu_1 = \mu_2 = \mu_3$
$H_1:$ At least one mean is different from the others. (claim)

C. V. = 3.55      $\alpha = 0.05$
d. f. N = 2      d. f. D = 18

$\overline{X}_1 = 29.625$      $s_1^2 = 59.125$
$\overline{X}_2 = 29$         $s_2^2 = 63.333$
$\overline{X}_3 = 28.5$       $s_3^2 = 37.1$

$\overline{X}_{GM} = 29.095$

$s_B^2 = \dfrac{\sum n_i(\overline{X}_i - \overline{X}_{GM})^2}{k-1}$

$s_B^2 = \dfrac{8(29.625 - 29.095)^2}{2} + \dfrac{7(29 - 29.095)^2}{2}$

$+ \dfrac{6(28.5 - 29.095)^2}{2} = 2.21726$

$s_W^2 = \dfrac{\sum(n_i - 1)s_i^2}{\sum(n_i - 1)}$

3. continued

$$s_W^2 = \frac{7(59.125)+6(63.333)+5(37.1)}{7+6+5}$$

$$s_W^2 = 54.509611$$

$$F = \frac{s_B^2}{s_W^2} = \frac{2.21726}{54.509611} = 0.04075$$

Do not reject the null hypothesis. There is not enough evidence to support the claim that at least one mean is different from the others.

5.
$H_0$: $\mu_1 = \mu_2 = \mu_3$
$H_1$: At least one mean is different. (claim)
C. V. = 2.61          $\alpha = 0.10$
d. f. N = 2          d. f. D = 19

$$\overline{X}_{GM} = 3.8591$$

$$s_B^2 = 1.65936$$

$$s_W^2 = 3.40287$$

$$F = \frac{1.65936}{3.40287} = 0.4876$$

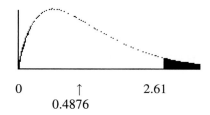

0          ↑          2.61
    0.4876

Do not reject. There is not enough evidence to support the claim that at least one mean is different from the others.

7.
$H_0$: $\mu_1 = \mu_2 = \mu_3 = \mu_4$
$H_1$: At least one mean is different from the others. (claim)

C. V. = 3.59          $\alpha = 0.05$
d. f. N = 3          d. f. D = 11

$$\overline{X}_{GM} = 12.267$$

$$s_B^2 = 21.422$$

$$s_W^2 = 117.697$$

$$F = \frac{21.422}{117.697} = 0.182$$

7. continued

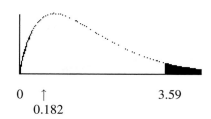

0          ↑                    3.59
    0.182

Do not reject the null hypothesis. There is not enough evidence to support the claim that at least one mean is different from the others.

9.
$H_0$: There is no interaction effect between the type of exercise program and the type of diet on a person's glucose level.
$H_1$: There is an interaction effect between the type of exercise program and the type of diet on a person's glucose level.

$H_0$: There is no difference in the means for the glucose levels of the persons in the two exercise programs.
$H_1$: There is a difference in the means for the glucose levels of the persons in the two exercise programs.

$H_0$: There is no difference in the means for the glucose levels of the persons in the two diet programs.
$H_1$: There is a difference in the means for the glucose levels of the persons in the two diet programs.

### ANOVA SUMMARY TABLE

| Source | SS | d. f. | MS | F |
|---|---|---|---|---|
| Exercise | 816.750 | 1 | 816.750 | 60.50 |
| Diet | 102.083 | 1 | 102.083 | 7.56 |
| Interaction | 444.083 | 1 | 444.083 | 32.90 |
| Within | 108.000 | 8 | 13.500 | |
| Total | 1470.916 | 11 | | |

At $\alpha = 0.05$ and d. f. N = 1 and d. f. D = 8 the critical value is 5.32 for each $F_A$, $F_B$, and $F_{AxB}$.

Hence all three null hypotheses are rejected.

9. continued

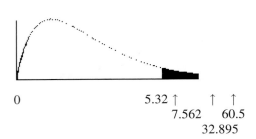

0                   5.32 ↑    ↑  ↑
                          7.562    60.5
                                32.895

The cell means should be calculated.

| Exercise | Diet A | Diet B |
|----------|--------|--------|
| I | 64.000 | 57.667 |
| II | 68.333 | 86.333 |

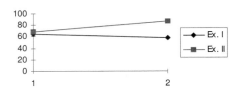

Since the means for the Exercise Program I are both smaller than those for Exercise Program II and the vertical differences are not the same, the interaction is ordinal. Hence one can say that there is a difference for exercise, diet; and that an interaction effect is present.

### CHAPTER 12 QUIZ

1. False, there could be a significant difference between only some of the means.
2. False, degrees of freedom are used to find the critical value.
3. False, the null hypothesis should not be rejected.
4. True
5. d.
6. a.
7. a.
8. c.
9. ANOVA
10. Tukey
11. two

12. $H_0$: $\mu_1 = \mu_2 = \mu_3 = \mu_4$
$H_1$: At least one mean is different from the others. (claim)

12. continued
C. V. $= 3.49$
$s_B^2 = 116.646$    $s_W^2 = 36.132$
$F = \frac{116.646}{36.132} = 3.23$

Do not reject the null hypothesis. There is not enough evidence to support the claim that the means are different.

13. $H_0$: $\mu_1 = \mu_2 = \mu_3$
$H_1$: At least one mean is different from the others. (claim)
C. V. $= 6.93$    $\alpha = 0.01$
$s_B^2 = 119.467$    $s_W^2 = 34.167$
$F = \frac{119.467}{34.167} = 3.497$

Do not reject the null hypothesis. There is not enough evidence to support the claim that at least one mean is different from the others. Writers would want to target their material to the age group of the viewers.

14. $H_0$: $\mu_1 = \mu_2 = \mu_3$
$H_1$: At least one mean is different from the others. (claim)
C. V. $= 3.68$    $\alpha = 0.05$
$s_B^2 = 617.167$    $s_W^2 = 58.811$
$F = \frac{617.167}{58.811} = 10.494$
Reject the null hypothesis. There is enough evidence to support the claim that at least one mean is different from the others.

Tukey Test:
C. V. $= 3.67$
$\overline{X}_1 = 47.67$
$\overline{X}_2 = 63$
$\overline{X}_3 = 43.83$
$\overline{X}_1$ vs $\overline{X}_2$: q $= -4.90$
$\overline{X}_1$ vs $\overline{X}_3$: q $= 1.23$
$\overline{X}_2$ vs $\overline{X}_3$: q $= 6.12$
There is a significant difference between $\overline{X}_1$ and $\overline{X}_2$ and between $\overline{X}_2$ and $\overline{X}_3$.

15. $H_0$: $\mu_1 = \mu_2 = \mu_3$ (claim)
$H_1$: At least one mean is different from the others.
C. V. $= 2.70$    $\alpha = 0.10$
$s_B^2 = 0.1213$    $s_W^2 = 2.3836$
$F = \frac{0.1213}{2.3836} = 0.0509$
Do not reject. There is not enough evidence to reject the claim that the means are the same.

16. $H_0$: $\mu_1 = \mu_2 = \mu_3 = \mu_4$
$H_1$: At least one mean is different from the others.  (claim)
C. V. $= 3.07$ $\qquad \alpha = 0.05$
$s_B^2 = 15.3016$ $\qquad s_W^2 = 33.5283$
$F = \frac{15.3016}{33.5283} = 0.4564$
Do not reject.  There is not enough evidence to support the claim that at least one mean is different.


17.
a.  two-way ANOVA
b.  diet and exercise program
c.  2

d.  $H_0$:  There is no interaction effect between the type of exercise program and the type of diet on a person's weight loss.
$H_1$:  There is an interaction effect between the type of exercise program and the type of diet on a person's weight loss.

$H_0$:  There is no difference in the means of the weight losses for those in the exercise programs.
$H_1$:  There is a difference in the means of the weight losses for those in the exercise programs.

$H_0$:  There is no difference in the means of the weight losses for those in the diet programs.
$H_1$:  There is a difference in the means of the weight losses for those in the diet programs.

e.  Diet:  $F = 21.0$, significant
Exercise Program:  $F = 0.429$, not significant
Interaction:  $F = 0.429$, not significant

f.  Reject the null hypothesis for the diets.

Note: Graphs are not to scale and are intended to convey a general idea. Answers may vary due to rounding.

## EXERCISE SET 13-1

**1.**
Non-parametric means hypotheses other than those using population parameters can be tested, whereas distribution free means no assumptions about the population distributions have to be satisfied.

**3.**
The advantages of non-parametric methods are:
1. They can be used to test population parameters when the variable is not normally distributed.
2. They can be used when data is nominal or ordinal in nature.
3. They can be used to test hypotheses other than those involving population parameters.
4. The computations are easier in some cases than the computations of the parametric counterparts.
5. They are easier to understand.

The disadvantages are:
1. They are less sensitive than their parametric counterparts.
2. They tend to use less information than their parametric counterparts.
3. They are less efficient than their parametric counterparts.

**5.**

| DATA | 22 | 32 | 34 | 43 | 43 | 65 | 66 | 71 |
|------|----|----|----|----|----|----|----|----|
| RANK | 1 | 2 | 3 | 4.5 | 4.5 | 6 | 7 | 8 |

**7.**

| DATA | 3.2 | 5.9 | 10.3 | 11.1 | 19.4 | 21.8 | 23.1 |
|------|-----|-----|------|------|------|------|------|
| RANK | 1 | 2 | 3 | 4 | 5 | 6 | 7 |

**9.**

| DATA | 11 | 28 | 36 | 41 | 47 | 50 | 50 | 50 |
|------|----|----|----|----|----|----|----|----|
| RANK | 1 | 2 | 3 | 4 | 5 | 7 | 7 | 7 |

| DATA | 52 | 71 | 71 | 88 |
|------|----|----|----|----|
| RANK | 9 | 10.5 | 10.5 | 12 |

## EXERCISE SET 13-2

**1.**
The sign test uses only + or − signs.

**3.**
The smaller number of + or − signs.

**5.**

| | | | | |
|--|--|--|--|--|
| − | − | + | + | − |
| 0 | + | − | + | + |
| + | − | + | − | + |
| + | − | + | − | + |

$H_0$: Median = 38
$H_1$: Median $\neq$ 38

$\alpha = 0.05$    n = 20
C. V. = 5
Test value = 8

Since $8 > 5$, do not reject the null hypothesis. There is not enough evidence to reject the claim that the median is 38.

**7.**

| | | | | |
|--|--|--|--|--|
| − | + | + | − | − |
| + | + | + | − | − |
| − | − | + | − | − |
| − | + | − | 0 | − |

$H_0$: Median = 25   (claim)
$H_1$: Median $\neq$ 25

$\alpha = 0.05$    n = 19
C. V. = 4    Test value = 7

Since $7 > 4$, do not reject the null hypothesis. There is not enough evidence to reject the claim that the median is 25. School boards could use the median to plan for the costs of cyber school enrollments.

**9.**
$H_0$: median = \$10.86    (claim)
$H_1$: median $\neq$ \$10.86

C. V. = $\pm 1.96$

$$z = \frac{(x + 0.5) - \left(\frac{n}{2}\right)}{\frac{\sqrt{n}}{2}} = \frac{(18 + 0.5) - \frac{42}{2}}{\frac{\sqrt{42}}{2}}$$

$$= \frac{-2.5}{3.24} = -0.77$$

9. continued

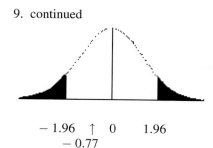

$$-1.96 \quad \uparrow \quad 0 \quad 1.96$$
$$-0.77$$

Do not reject the null hypothesis. There is not enough evidence to reject the claim that the median is $10.86. Home buyers could estimate the yearly cost of their gas bills.

11.
$H_0$: median = 150 (claim)
$H_1$: median $\neq$ 150

$$z = \frac{(x+0.5)-\left(\frac{n}{2}\right)}{\frac{\sqrt{n}}{2}} = \frac{(9+0.5)-\frac{35}{2}}{\frac{\sqrt{35}}{2}}$$

$$z = -2.70$$

C. V. $= \pm 1.96$

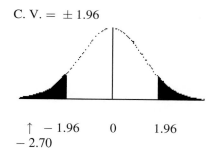

$$\uparrow \quad -1.96 \quad 0 \quad 1.96$$
$$-2.70$$

Reject. There is not enough evidence to support the claim that the median number of faculty members is 150.

13.
$H_0$: Median = 50 (claim)
$H_1$: Median $\neq$ 50

$$z = \frac{(x+0.5)-\left(\frac{n}{2}\right)}{\frac{\sqrt{n}}{2}} = \frac{(38+0.5)-\frac{100}{2}}{\frac{\sqrt{100}}{2}}$$

$$z = -2.3$$

P-value = 0.0214

Reject. There is enough evidence to reject the claim that 50% of the students are against extending the school year.

15.

| A | B | C | D | E | F | G | H |
|---|---|---|---|---|---|---|---|
| + | + | + | + | + | − | + | + |

$H_0$: The medication has no effect on weight loss.
$H_1$: The medication affects weight loss. (claim)
$\alpha = 0.05$    n = 8
C. V. = 0    Test value = 1

Since 1 > 0, do not reject the null hypothesis. There is not enough evidence to support the claim that the medication affects weight loss.

17.

| 1 | 2 | 3 | 4 | 5 | 6 | 7 | 8 |
|---|---|---|---|---|---|---|---|
| + | + | + | − | + | + | + | + |

$H_0$: There is no difference in attendance. (claim)
$H_1$: There is a difference in attendance.

$\alpha = 0.10$    n = 8
C. V. = 1    Test value = 1

Since the test value is equal to the critical value, reject the null hypothesis. There is enough evidence to reject the claim that there is no difference in attendance.

19.

| 1 | 2 | 3 | 4 | 5 | 6 | 7 | 8 | 9 | 10 |
|---|---|---|---|---|---|---|---|---|----|
| + | + | + | + | + | + | − | + | + | − |

$H_0$: The number of viewers is the same as last year. (claim)
$H_1$: The number of viewers is not the same as last year.

$\alpha = 0.01$    n = 10
C. V. = 0    Test value = 2

Since 2 > 1, do not reject the null hypothesis. There is not enough evidence to reject the claim that the number of viewers is the same as last year.

21.
3, 4, 6, 9, 12, 15, 15, 16, 18, 22, 25, 30
At $\alpha = 0.05$, the value from Table J with n = 12 is 2; hence, count in 3 numbers from each end to get $6 \leq MD \leq 22$.

23.
4.2, 4.5, 4.7, 4.8, 5.1, 5.2, 5.6, 6.3, 7.1, 7.2,
7.8, 8.2, 9.3, 9.3, 9.5, 9.6

At $\alpha = 0.02$, the value from Table J with n =
16 is 2; hence, count 3 numbers from each
end to get $4.7 \leq MD \leq 9.3$

25.
12, 14, 14, 15, 16, 17, 18, 19, 19, 21, 23, 25,
27, 32, 33, 35, 39, 41, 42, 47
At $\alpha = 0.05$, the value from Table J with
n = 20 is 5; hence, count in 6 numbers from
each end to get $17 \leq MD \leq 33$.

EXERCISE SET 13-3

1.
The sample sizes must be greater than or
equal to 10.

3.
The standard normal distribution.

5.
$H_0$: There is no difference in calories.
$H_1$: There is a difference in calories between
the two delis. (claim)

C. V. = $\pm 1.96$    $\alpha = 0.05$

| 400 | 420 | 420 | 430 | 430 | 450 | 480 | 570 | 590 |
|-----|-----|-----|-----|-----|-----|-----|-----|-----|
| 1 | 2.5 | 2.5 | 4.5 | 4.5 | 6 | 7 | 8 | 9 |
| B | A | A | B | B | B | A | A | A |

| 610 | 620 | 630 | 680 | 690 | 710 | 740 | 750 | 760 |
|-----|-----|-----|-----|-----|-----|-----|-----|-----|
| 10 | 11 | 12 | 13 | 14 | 15 | 16 | 17 | 18 |
| A | A | A | B | B | B | A | B | B |

| 790 | 860 |
|-----|-----|
| 19 | 20 |
| A | B |

R = 97

$$\mu_R = \frac{n_1(n_1 + n_2 + 1)}{2}$$

$$\mu_R = \frac{10(10 + 10 + 1)}{2} = \frac{10(21)}{2} = \frac{210}{2} = 105$$

$$\sigma_R = \sqrt{\frac{n_1 \cdot n_2(n_1 + n_2 + 1)}{12}}$$

$$\sigma_R = \sqrt{\frac{10 \cdot 10(10 + 10 + 1)}{12}} = \sqrt{\frac{(10)(10)(21)}{12}}$$

$$\sigma_R = \sqrt{175} = 13.23$$

5. continued
$$Z = \frac{R - \mu_R}{\sigma_R} = \frac{97 - 105}{13.23} = -0.60$$

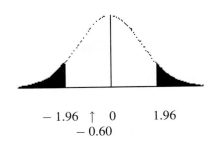

$-1.96 \uparrow 0$     $1.96$
$-0.60$

Do not reject the null hypothesis. There is
not enough evidence to support the claim
that there is a difference in calories.

7.
$H_0$: There is no difference in the stopping
distances of the two types of automobiles.
(claim)
$H_1$: There is a difference between the
stopping distances of the two types of
automobiles.

C. V. = $\pm 1.65$

| 186 | 187 | 188 | 188 | 190 | 192 | 193 | 194 | 195 | 196 |
|-----|-----|-----|-----|-----|-----|-----|-----|-----|-----|
| 1 | 2 | 3.5 | 3.5 | 5 | 6 | 7 | 8 | 9 | 10 |
| M | M | M | M | M | M | C | M | M | C |

| 198 | 200 | 203 | 204 | 206 | 211 | 212 | 214 | 218 | 297 |
|-----|-----|-----|-----|-----|-----|-----|-----|-----|-----|
| 11 | 12 | 13 | 14 | 15 | 16 | 17 | 18 | 19 | 20 |
| C | C | M | C | C | C | C | M | C | C |

R = 69

$$\mu_R = \frac{10(10 + 10 + 1)}{2} = 105$$

$$\sigma_R = \sqrt{\frac{10 \cdot 10(10 + 10 + 1)}{12}} = 13.23$$

$$z = \frac{69 - 105}{13.23} = -2.72$$

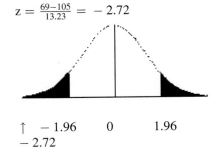

$\uparrow \; -1.96$     $0$     $1.96$
$-2.72$

**7. continued**

Reject the null hypothesis. There is enough evidence to reject the claim that there is no difference in the stopping distances of the two types of automobiles.

**9.**

$H_0$: There is no difference in the number of hunting accidents in the two regions.
$H_1$: There is a difference in the number of hunting accidents in the two regions. (claim)

C. V. $= \pm 1.96$

| 2 | 3 | 5 | 5 | 6 | 7 | 8 | 8 | 9 | 10 |
|---|---|---|---|---|---|---|---|---|---|
| 1 | 2 | 3.5 | 3.5 | 5 | 6 | 7.5 | 7.5 | 9 | 10 |
| E | E | E | E | E | E | E | W | W | W |

| 11 | 11 | 11 | 13 | 13 | 14 | 15 | 17 | 17 | 21 |
|---|---|---|---|---|---|---|---|---|---|
| 12 | 12 | 12 | 14.5 | 14.5 | 16 | 17 | 18.5 | 18.5 | 20 |
| W | W | E | E | W | E | W | W | W | W |

$R = 71$

$$\mu_R = \frac{n_1(n_1 + n_2 + 1)}{2} = \frac{10(10 + 10 + 1)}{2} = 105$$

$$\sigma_R = \sqrt{\frac{n_1 \cdot n_2(n_1 + n_2 + 1)}{12}}$$

$$= \sqrt{\frac{10 \cdot 10(10 + 10 + 1)}{12}} = 13.23$$

$$Z = \frac{R - \mu_R}{\sigma_R} = \frac{71 - 105}{13.23} = -2.57$$

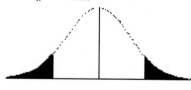

↑ $-1.96$    0    1.96
$-2.57$

Reject the null hypothesis. There is enough evidence to support the claim that there is a difference in the number of accidents in the two areas. The number of accidents may be related to the number of hunters in the two areas.

**11.**

$H_0$: There is no difference in the times needed to assemble the product.
$H_1$: There is a difference in the times needed to assemble the product. (claim)
C. V. $= \pm 1.96$

**11. continued**

| 1.6 | 1.7 | 1.9 | 2.0 | 2.4 | 2.6 | 2.7 | 2.9 | 3.0 |
|---|---|---|---|---|---|---|---|---|
| 1 | 2 | 3 | 4 | 5 | 6 | 7 | 8 | 9 |
| N | N | N | N | N | N | N | N | G |

| 3.1 | 3.2 | 3.4 | 3.6 | 3.8 | 3.9 | 4.2 | 4.4 | 4.7 |
|---|---|---|---|---|---|---|---|---|
| 10 | 11 | 12 | 13 | 14 | 15 | 16 | 17 | 18 |
| N | G | N | G | N | N | G | G | G |

| 5.3 | 5.6 | 5.8 | 6.3 | 6.4 | 7.1 | 7.3 | 8.2 |
|---|---|---|---|---|---|---|---|
| 19 | 20 | 21 | 22 | 23 | 24 | 25 | 26 |
| N | G | G | G | G | G | G | G |

$R = 245$

$$\mu_R = \frac{n_1(n_1 + n_2 + 1)}{2} = \frac{13(13 + 13 + 1)}{2}$$

$$\mu_R = \frac{13(27)}{2} = 175.5$$

$$\sigma_R = \sqrt{\frac{n_1 \cdot n_2(n_1 + n_2 + 1)}{12}}$$

$$\sigma_R = \sqrt{\frac{13 \cdot 13(13 + 13 + 1)}{12}} = 19.5$$

$$Z = \frac{R - \mu_R}{\sigma_R} = \frac{245 - 175.5}{19.5} = \frac{69.5}{19.5} = 3.56$$

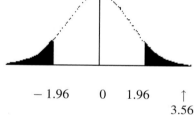

$-1.96$    0    1.96    ↑
                               3.56

Reject the null hypothesis. There is enough evidence to support the claim that there is a difference in the productivity of the two groups.

**EXERCISE SET 13-4**

**1.**

The t-test for dependent samples.

**3.**

| B | A | B − A | \|B − A\| | Rank | Signed Rank |
|---|---|---|---|---|---|
| 108 | 110 | − 2 | 2 | 1 | − 1 |
| 97 | 97 | 0 | | | |
| 115 | 103 | 12 | 12 | 4.5 | 4.5 |
| 162 | 168 | − 6 | 6 | 2 | − 2 |
| 156 | 143 | 13 | 13 | 6 | 6 |
| 105 | 112 | − 7 | 7 | 3 | − 3 |
| 153 | 141 | 12 | 12 | 4.5 | 4.5 |

3. continued

Sum of the $-$ ranks: $-6$.

Sum of the + ranks: 15

$w_s = 6$

5.

C. V. = 20        $w_s = 18$

Since $18 \leq 20$, reject the null hypothesis.

7.

C. V. = 102        $w_s = 102$

Since $102 = 102$, reject the null hypothesis.

9.

$H_0$: The human dose is less than or equal to the animal dose.

$H_1$: The human dose is more than the animal dose. (claim)

| H | A | H − A | \|H − A\| | Rank | Signed Rank |
|---|---|---|---|---|---|
| 0.67 | 0.13 | 0.54 | 0.54 | 7 | + 7 |
| 0.64 | 0.18 | 0.46 | 0.46 | 6 | + 6 |
| 1.20 | 0.42 | 0.78 | 0.78 | 8 | + 8 |
| 0.51 | 0.25 | 0.26 | 0.26 | 4 | + 4 |
| 0.87 | 0.57 | 0.30 | 0.30 | 5 | + 5 |
| 0.74 | 0.57 | 0.17 | 0.17 | 3 | + 3 |
| 0.50 | 0.49 | 0.01 | 0.01 | 1 | + 1 |
| 1.22 | 1.28 | − 0.06 | 0.06 | 2 | − 2 |

Sum of the $-$ ranks: $-2$

Sum of the + ranks: $+34$

$n = 8$    C. V. = 6

$w_s = |-2| = 2$

Since $2 < 6$, reject the null hypothesis. There is enough evidence to support the claim that the human dose costs more than the animal dose. One reason is that some people might not be inclined to pay a lot of money for their pets' medication.

11.

$H_0$: There is no difference in test scores.

$H_1$: There is a difference in test scores.

$n = 9$    $\alpha = 0.05$    C. V. = 6

11. continued

| 1 | 2 | 1 − 2 | \|1 − 2\| | Rank | Signed Rank |
|---|---|---|---|---|---|
| 78 | 85 | − 7 | 7 | 7 | − 7 |
| 95 | 92 | 3 | 3 | 3.5 | 3.5 |
| 72 | 70 | 2 | 2 | 2 | 2 |
| 65 | 68 | − 3 | 3 | 3.5 | − 3.5 |
| 70 | 69 | 1 | 1 | 1 | 1 |
| 70 | 76 | − 6 | 6 | 6 | − 6 |
| 79 | 88 | − 9 | 9 | 8 | − 8 |
| 85 | 96 | − 11 | 11 | 9 | − 9 |
| 75 | 80 | − 5 | 5 | 5 | − 5 |

Sum of the $-$ ranks: $-38.5$

Sum of the + ranks: 6.5

$w_s = 6.5$

Since $w_s >$ C. V., do not reject the null hypothesis. There is not enough evidence to conclude that there is a difference in test scores.

13.

$H_0$: The prices of prescription drugs in the United States are greater than or equal to the prices in Canada.

$H_1$: The drugs sold in Canada are cheaper. (claim)

$n = 10$    $\alpha = 0.05$    C. V. = 11

| U. S. | C | US − C | \|US − C\| | Rank | Signed Rank |
|---|---|---|---|---|---|
| 3.31 | 1.47 | 1.84 | 1.84 | 8 | + 8 |
| 2.27 | 1.07 | 1.20 | 1.20 | 4.5 | + 4.5 |
| 2.54 | 1.34 | 1.20 | 1.20 | 4.5 | + 4.5 |
| 3.13 | 1.34 | 1.79 | 1.79 | 7 | + 7 |
| 23.40 | 21.44 | 1.94 | 1.94 | 10 | + 10 |
| 3.16 | 1.47 | 1.69 | 1.69 | 6 | + 6 |
| 1.98 | 1.07 | 0.91 | 0.91 | 3 | + 3 |
| 5.27 | 3.39 | 1.88 | 1.88 | 9 | + 9 |
| 1.96 | 2.22 | − 0.26 | 0.26 | 2 | − 2 |
| 1.11 | 1.13 | − 0.02 | 0.02 | 1 | − 1 |

Sum of the + ranks:

$8 + 4.5 + 4.5 + 7 + 10 + 6 + 3 + 9 = 52$

Sum of the − ranks: $(-2) + (-1) = -3$

$w_s = |-3| = 3$

Since $3 < 11$, reject the null hypothesis. There is enough evidence to support the claim that the drugs are less expensive in Canada.

EXERCISE SET 13-5

**1.**

$H_0$: There is no difference in the number of calories each brand contains.

$H_1$: There is a difference in the number of calories each brand contains. (claim)

C. V. = 7.815    $\alpha = 0.05$    d. f. = 3

| A | Rank | B | Rank | C | Rank | D | Rank |
|---|------|---|------|---|------|---|------|
| 112 | 7 | 110 | 6 | 109 | 5 | 106 | 3 |
| 120 | 13 | 118 | 12 | 116 | 9.5 | 122 | 15 |
| 135 | 24 | 123 | 16 | 125 | 17.5 | 130 | 21.5 |
| 125 | 17.5 | 128 | 19.5 | 130 | 21.5 | 117 | 11 |
| 108 | 4 | 102 | 2 | 128 | 19.5 | 116 | 9.5 |
| 121 | 14 | 101 | 1 | 132 | 23 | 114 | 8 |
| $R_1=$ | 79.5 | $R_2=$ | 56.5 | $R_3=$ | 96 | $R_4=$ | 68 |

$$H = \frac{12}{N(N+1)} \left( \frac{R_1^2}{n_1} + \frac{R_2^2}{n_2} + \frac{R_3^2}{n_3} + \frac{R_4^2}{n_4} \right) - 3(N+1)$$

$$H = \frac{12}{12(24+1)} \left( \frac{79.5^2}{6} + \frac{56.5^2}{6} + \frac{96^2}{6} + \frac{68^2}{6} \right)$$

$$H = 2.842$$

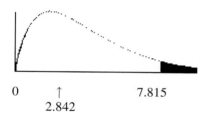

0        ↑        7.815
       2.842

Do not reject the null hypothesis. There is not enough evidence to support the claim that there is a difference in the calories.

**3.**

$H_0$: There is no difference in the prices of the three types of lawnmowers.

$H_1$: There is a difference in the prices of the three types of lawnmowers. (claim)

C. V. = 4.605    d. f. = 2    $\alpha = 0.10$

| Self | Rank | Push | Rank | Electric | Rank |
|------|------|------|------|----------|------|
| 290 | 7 | 320 | 9 | 188 | 2 |
| 325 | 10 | 360 | 12 | 245 | 6 |
| 210 | 4 | 200 | 3 | 470 | 14 |
| 300 | 8 | 229 | 5 | 395 | 13 |
| 330 | 11 | 160 | 1 | | |
| $R_1=$ | 40 | $R_2=$ | 30 | $R_3=$ | 35 |

$$H = \frac{12}{N(N+1)} \left( \frac{R_1^2}{n_1} + \frac{R_2^2}{n_2} + \frac{R_3^2}{n_3} \right)$$

$$- 3(N+1)$$

**3. continued**

$$H = \frac{12}{14(14+1)} \left( \frac{40^2}{5} + \frac{30^2}{5} + \frac{35^2}{4} \right) - 3(14+1)$$

$$H = \frac{12}{210} (320 + 180 + 306.25) - 3(15)$$

$$H = 1.07$$

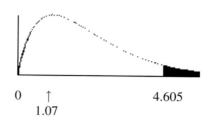

0     ↑                    4.605
    1.07

Do not reject the null hypothesis. There is not enough evidence to support the claim that the prices are different. No, price is not a factor. Results are suspect since one sample is less than 5.

**5.**

$H_0$: There is no difference in the number of carbohydrates in one serving of each of the three types of food.

$H_1$: There is a difference in the number of carbohydrates in each of the three types of food. (claim)

C. V. = 9.210    d. f. = 2    $\alpha = 0.01$

| Pasta | Rank | Ice Cream | Rank | Bread | Rank |
|-------|------|-----------|------|-------|------|
| 11 | 4 | 5 | 1 | 43 | 11 |
| 22 | 8 | 13 | 6 | 62 | 14 |
| 16 | 7 | 10 | 3 | 71 | 15 |
| 29 | 10 | 8 | 2 | 49 | 12 |
| 25 | 9 | 12 | 5 | 50 | 13 |
| $R_1=$ | 38 | $R_2=$ | 17 | $R_3=$ | 65 |

$$H = \frac{12}{N(N+1)} \left( \frac{R_1^2}{n_1} + \frac{R_2^2}{n_2} + \frac{R_3^2}{n_3} + \frac{R_4^2}{n_4} \right) - 3(N+1)$$

$$= \frac{12}{15(15+1)} \left( \frac{38^2}{5} + \frac{17^2}{5} + \frac{65^2}{5} \right) - 3(15+1)$$

$$H = 11.58$$

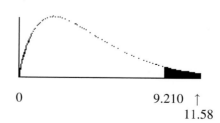

0                        9.210  ↑
                               11.58

5. continued

Reject the null hypothesis. There is enough evidence to support the claim that there is a difference in the number of carbohydrates in the three foods. You should recommend the ice cream.

7.

$H_0$: There is no difference in spending between regions.

$H_1$: There is a difference in spending between regions.

C. V. = 5.991

| E | Rank | M | Rank | W | Rank |
|---|---|---|---|---|---|
| 6701 | 3 | 9854 | 15 | 7584 | 10 |
| 6708 | 4 | 8414 | 11 | 5474 | 1 |
| 9186 | 12 | 7279 | 7 | 6622 | 2 |
| 6786 | 5 | 7311 | 8 | 9673 | 14 |
| 9261 | 13 | 6947 | 6 | 7353 | 9 |
| $R_1=$ | 37 | $R_2=$ | 47 | $R_3=$ | 36 |

$$H = \frac{12}{N(N+1)} \left( \frac{R_1^2}{n_1} + \frac{R_2^2}{n_2} + \frac{R_3^2}{n_3} + \frac{R_4^2}{n_4} \right) - 3(N+1)$$

$$H = \frac{12}{15(15+1)} \left( \frac{37^2}{5} + \frac{47^2}{5} + \frac{36^2}{5} \right) - 3(15+1)$$

$$H = 0.74$$

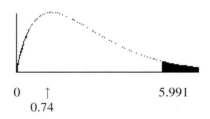

0    ↑         5.991
    0.74

Do not reject the null hypothesis. There is not enough evidence to conclude that there is a difference in spending.

9.

$H_0$: There is no difference in the number of crimes in the 5 precincts.

$H_1$: There is a difference in the number of crimes in the 5 precincts. (claim)

C. V. = 13.277    d. f. = 4    $\alpha = 0.01$

| 1 | Rank | 2 | Rank | 3 | Rank |
|---|---|---|---|---|---|
| 105 | 24 | 87 | 13 | 74 | 7.5 |
| 108 | 25 | 86 | 12 | 83 | 11 |
| 99 | 22 | 91 | 16 | 78 | 9 |
| 97 | 20 | 93 | 18 | 74 | 7.5 |
| 92 | 17 | 82 | 10 | 60 | 5 |
| $R_1=$ | 108 | $R_2=$ | 69 | $R_3=$ | 40 |

9. continued

| 4 | Rank | 5 | Rank |
|---|---|---|---|
| 56 | 3 | 103 | 23 |
| 43 | 1 | 98 | 21 |
| 52 | 2 | 94 | 19 |
| 58 | 4 | 89 | 15 |
| 62 | 6 | 88 | 14 |
| $R_4=$ | 16 | $R_5=$ | 92 |

$$H = \frac{12}{N(N+1)} \left( \frac{R_1^2}{n_1} + \frac{R_2^2}{n_2} + \frac{R_3^2}{n_3} + \frac{R_4^2}{n_4} + \frac{R_5^2}{n_5} \right) - 3(N+1)$$

$$H = \frac{12}{25(25+1)} \left( \frac{108^2}{5} + \frac{69^2}{5} + \frac{40^2}{5} + \frac{16^2}{5} + \frac{92^2}{5} \right) - 3(25+1)$$

$$H = 20.753$$

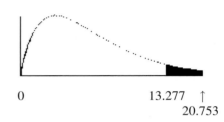

0             13.277   ↑
                 20.753

Reject the null hypothesis. There is enough evidence to support the claim that there is a difference in the number of crimes for the precincts.

11.

$H_0$: There is no difference in speeds.

$H_1$: There is a difference in speeds.

d. f. = 2    $\alpha = 0.10$

C. V. = 5.991

| Pred. | Rank | Deer | Rank | Dom. | Rank |
|---|---|---|---|---|---|
| 70 | 15 | 50 | 12.5 | 47.5 | 11 |
| 50 | 12.5 | 35 | 5.5 | 39.35 | 7 |
| 43 | 10 | 32 | 4 | 35 | 5.5 |
| 42 | 9 | 30 | 2.5 | 30 | 2.5 |
| 40 | 8 | 61 | 14 | 11 | 1 |
| $R_1=$ | 54.5 | $R_2=$ | 38.5 | $R_3=$ | 27 |

$$H = \frac{12}{N(N+1)} \left( \frac{R_1^2}{n_1} + \frac{R_2^2}{n_2} + \frac{R_3^2}{n_3} \right) - 3(N+1)$$

$$H = \frac{12}{12(15+1)} \left( \frac{54.5^2}{5} + \frac{38.5^2}{5} + \frac{27^2}{5} \right) - 3(15+1)$$

$$H = 3.815$$

Do not reject. There is not enough evidence to conclude there is a difference in speeds.

EXERCISE SET 13-6

**1.**
0.716

**3.**
0.648

**5.**
$H_0$: $\rho = 0$
$H_1$: $\rho \neq 0$
C. V. $= \pm 0.564$    n $= 10$    $\alpha = 0.10$

| Tornadoes | $R_1$ | Temp | $R_2$ | $R_1 - R_2$ | $d^2$ |
|---|---|---|---|---|---|
| 668 | 6 | 112 | 5 | 1 | 1 |
| 781 | 7 | 118 | 9 | $-2$ | 4 |
| 1590 | 10 | 109 | 3 | 7 | 49 |
| 798 | 8 | 117 | 8 | 0 | 0 |
| 1198 | 9 | 121 | 10 | $-1$ | 1 |
| 169 | 2 | 108 | 2 | 0 | 0 |
| 310 | 3 | 111 | 4 | $-1$ | 1 |
| 360 | 4 | 113 | 6 | $-2$ | 4 |
| 21 | 1 | 105 | 1 | 0 | 0 |
| 625 | 5 | 114 | 7 | $-2$ | 4 |
| | | | | $\sum d^2 =$ | 64 |

$r_s = 1 - \frac{6 \cdot \sum d^2}{n(n^2-1)} = 1 - \frac{6 \cdot 64}{10(10^2-1)}$
$r_s = 0.612$

Reject the null hypothesis. There is a significant relationship between the number of tornadoes and high temperatures.

**7.**
$H_0$: $\rho = 0$
$H_1$: $\rho \neq 0$

C. V. $= \pm 0.700$    n $= 9$    $\alpha = 0.05$

| Releases | $R_1$ | Receipts | $R_2$ | $R_1 - R_2$ | $d^2$ |
|---|---|---|---|---|---|
| 361 | 9 | 3844 | 9 | 0 | 0 |
| 270 | 7 | 1967 | 8 | $-1$ | 1 |
| 306 | 8 | 1371 | 7 | 1 | 1 |
| 22 | 5 | 1064 | 6 | $-1$ | 1 |
| 35 | 6 | 667 | 5 | 1 | 1 |
| 10 | 2 | 241 | 4 | $-2$ | 4 |
| 8 | 1 | 188 | 3 | $-2$ | 4 |
| 12 | 3 | 154 | 2 | 1 | 1 |
| 21 | 4 | 125 | 1 | 3 | 9 |
| | | | | $\sum d^2 =$ | 22 |

$r_s = 1 - \frac{6 \cdot \sum d^2}{n(n^2-1)}$
$r_s = 1 - \frac{6 \cdot 22}{9(9^2-1)} = 0.817$

**7. continued**
Reject. There is a significant relationship between the number of new releases and gross receipts.

**9.**
$H_0$: $\rho = 0$
$H_1$: $\rho \neq 0$

C. V. $= \pm 0.738$

| Teen | $R_1$ | Parent | $R_2$ | $R_1 - R_2$ | $d^2$ |
|---|---|---|---|---|---|
| 4 | 4 | 1 | 1 | 3 | 9 |
| 6 | 6 | 7 | 7 | $-1$ | 1 |
| 2 | 2 | 5 | 5 | $-3$ | 9 |
| 8 | 8 | 4 | 4 | 4 | 16 |
| 1 | 1 | 3 | 3 | $-2$ | 4 |
| 7 | 7 | 8 | 8 | $-1$ | 1 |
| 3 | 3 | 2 | 2 | 1 | 1 |
| 5 | 5 | 6 | 6 | $-1$ | 3 |
| | | | | $\sum d^2 =$ | 42 |

$r_s = 1 - \frac{6 \cdot 42}{8(8^2-1)} = 1 - \frac{252}{504} = 0.5$

Do not reject the null hypothesis. There is not enough evidence to say that a significant relationship exists between the rankings.

**11.**
$H_0$: $\rho = 0$
$H_1$: $\rho \neq 0$
C. V. $= \pm 0.700$

| Agency | $R_1$ | Station | $R_2$ | $R_1 - R_2$ | $d^2$ |
|---|---|---|---|---|---|
| 5.12 | 1.5 | 2.09 | 3 | $-1.5$ | 2.25 |
| 5.27 | 5 | 1.96 | 2 | 3.0 | 9.00 |
| 5.29 | 6 | 2.29 | 9 | $-3.0$ | 9.00 |
| 5.18 | 4 | 1.94 | 1 | 3.0 | 9.00 |
| 5.59 | 9 | 2.20 | 7.5 | 1.5 | 2.25 |
| 5.30 | 7 | 2.20 | 7.5 | $-0.5$ | 0.25 |
| 5.83 | 10 | 2.40 | 10 | 0 | 0 |
| 5.46 | 8 | 2.12 | 5 | 3.0 | 9.00 |
| 5.12 | 1.5 | 2.15 | 6 | $-4.5$ | 20.25 |
| 5.15 | 3 | 2.11 | 4 | $-1.0$ | 1.00 |
| | | | | $\sum d^2 =$ | 62.00 |

$r_s = 1 - \frac{6 \sum d^2}{n(n^2-1)} = 1 - \frac{6 \cdot 62}{10(10^2-1)} = 0.624$

Do not reject the null hypothesis. There is no significant linear relationship between gasoline prices at a car rental agency and prices at a gas station. One would wonder how the car rental agencies determine their prices.

13.

$H_0$: $\rho = 0$
$H_1$: $\rho \neq 0$

C. V. = $\pm 0.900$

| Students | $R_1$ | Cost | $R_2$ | $R_1 - R_2$ | $d^2$ |
|----------|-------|------|-------|-------------|-------|
| 10 | 3 | 7200 | 2 | 1 | 1 |
| 6 | 1 | 9393 | 5 | $-4$ | 16 |
| 17 | 5 | 7385 | 3 | 2 | 4 |
| 8 | 2 | 4500 | 1 | 1 | 1 |
| 11 | 4 | 8203 | 4 | 0 | 0 |
| | | | | $\sum d^2 =$ | 22 |

$r_s = 1 - \frac{6\sum d^2}{n(n^2-1)} = 1 - \frac{6 \cdot 22}{5(5^2-1)}$

$r_s = -0.10$

Do not reject the null hypothesis. There is no significant relationship exists between the number of cyber school students and the cost per pupil. In this case, the cost per pupil is different in each district.

15.

$H_0$: The occurrance of cavities is random.
$H_1$: The null hypothesis is not true.

The median of the data set is two. Using A = above and B = below, the runs (going across) are shown:

B AA B AAA B A BB AAAA B A B A B A
B A B AAA B A BB

There are 21 runs. The expected number of runs is between 10 and 22. Therefore, the null hypothesis should not be rejected. The number of cavities occurs at random.

17.

$H_0$: The lotto numbers occur at random.
$H_1$: The null hypothesis is not true.

OO E OO EE O EE OOO EE OO E O E O
EE

There are 14 runs and this value is between 7 and 18. Hence, do not reject the null hypothesis. The numbers occur at random.

19.

$H_0$: The seating occurs at random.
$H_1$: The null hypothesis is not true.

19. continued
BB GG BB G BBBBBB G BB GG BBBB
GGGG B G BBB GG

There are 14 runs and since this is between 10 and 23, the null hypothesis is not rejected. The seating occurs at random.

21.

$H_0$: The number of absences of employees occurs at random.
$H_1$: The null hypothesis is not true.

The median of the data is 12. Using A = above and B = below, the runs are shown.

A B AAAAAAA BBBBBBBBB AAAAAA
BBBB

There are six runs. The expected number of runs is between 9 and 21, hence the null hypothesis is rejected since six is not between 9 and 21. The number of absences do not occur at random.

23.
Answers will vary.

25.
$r = \frac{\pm z}{\sqrt{n-1}} = \frac{\pm 2.58}{\sqrt{30-1}} = \pm 0.479$

27.
$r = \frac{\pm z}{\sqrt{n-1}} = \frac{\pm 1.65}{\sqrt{60-1}} = \pm 0.215$

REVIEW EXERCISES - CHAPTER 13

1.
$H_0$: median $= 36$
$H_1$: median $\neq 36$

C. V. = $\pm 1.96$   $\alpha = 0.05$

$Z = \frac{(13+0.5) - \left(\frac{30}{2}\right)}{\frac{\sqrt{30}}{2}} = -0.548$

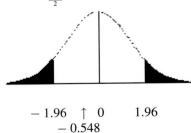

$-1.96$   $\uparrow$   $0$      $1.96$
$-0.548$

**1. continued**
Do not reject. There is not enough evidence to conclude that the median differs from 36.

**3.**
$H_0$: The special diet has no effect on weight.
$H_1$: The diet increases weight. (claim)
$R^+ = 35, R^- = 1$
C. V. = 0
Test value = 1

Do not reject the null hypothesis. There is not enough evidence to conclude a difference in prices. Comments: Examine what affects the result of this test.

**5.**
$H_0$: There is no difference in the amount of hours worked.
$H_1$: There is a difference in the amount of hours worked.

C. V. = $\pm 1.645$ at $\alpha = 0.10$

| 12 | 15 | 17 | 18 | 18 | 19 | 19 | 20 | 20 |
|----|----|----|----|----|----|----|----|----|
| 1 | 2 | 3 | 4.5 | 4.5 | 6.5 | 6.5 | 8.5 | 8.5 |
| A | A | A | A | L | A | L | A | L |

| 21 | 22 | 22 | 24 | 24 | 25 | 25 | 26 |
|----|----|----|----|----|----|----|----|
| 10 | 11.5 | 11.5 | 13.5 | 13.5 | 15.5 | 15.5 | 17 |
| L | A | L | A | L | L | A | L |

| 28 | 30 | 31 | 35 |
|----|----|----|----|
| 18 | 19 | 20 | 21 |
| L | A | L | L |

R = 85

$\mu_R = \frac{n_1(n_1 + n_2 + 1)}{2} = \frac{10(10 + 11 + 1)}{2} = 110$

$\sigma_R = \sqrt{\frac{n_1 n_2(n_1 + n_2 + 1)}{12}}$

$\sigma_R = \sqrt{\frac{10 \cdot 11(10 + 11 + 1)}{12}} = 14.201$

$Z = \frac{R - \mu_R}{\sigma_R} = \frac{85 - 110}{14.201} = -1.76$

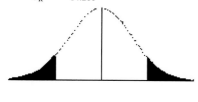

↑ $-1.65$    0    1.65
$-1.76$

**7. continued**
Reject the null hypothesis. There is enough evidence to conclude a difference in the hours worked.

At $\alpha = 0.05$, C. V. = $\pm 1.96$. The decision would be do not reject the null hypothesis.

**7.**
$H_0$: There is no difference in the amount spent.
$H_1$: There is a difference in the amount spent.

| B | A | B − A | \|B − A\| | Rank | Signed Rank |
|----|----|----|----|----|----|
| 7 | 6 | 1 | 1 | 1 | 1 |
| 5.5 | 10 | − 4.5 | 4.5 | 7 | − 7 |
| 4.5 | 7 | − 2.5 | 2.5 | 6 | − 6 |
| 10 | 12 | − 2 | 2 | 4 | − 4 |
| 6.75 | 8.5 | − 1.75 | 1.75 | 2 | − 2 |
| 5 | 7 | − 2 | 2 | 4 | − 4 |
| 6 | 8 | − 2 | 2 | 4 | − 4 |

Sum of the + ranks: 1
Sum of the − ranks: − 27
$w_s = 1$
C. V. = 2    $\alpha = 0.05$    n = 7

Reject the null hypothesis. There is enough evidence to conclude that there is a difference in amount spent.

**9.**
$H_0$: There is no difference in beach temperatures.
$H_1$: There is a difference in beach temperatures.

C. V. = 7.815

| S. Pac. | Rank | W. Gulf | Rank | E. Gulf | Rank |
|----|----|----|----|----|----|
| 67 | 4 | 86 | 21 | 87 | 25 |
| 68 | 5 | 86 | 21 | 87 | 25 |
| 66 | 3 | 84 | 13 | 86 | 21 |
| 69 | 6 | 85 | 16.5 | 86 | 21 |
| 63 | 2 | 79 | 8 | 85 | 16.5 |
| 62 | 1 | 85 | 16.5 | 84 | 13 |
| | | | | 85 | 16.5 |
| $R_1=$ | 21 | $R_2=$ | 96 | $R_3=$ | 138 |

| S. Atl. | Rank |
|----|----|
| 76 | 7 |
| 81 | 10 |
| 82 | 11 |
| 84 | 13 |
| 80 | 9 |
| 86 | 21 |
| 87 | 25 |
| $R_4=$ | 96 |

9. continued

$$H = \frac{12}{N(N+1)}\left(\frac{R_1^2}{n_1} + \frac{R_2^2}{n_2} + \frac{R_3^2}{n_3}\right) - 3(N+1)$$

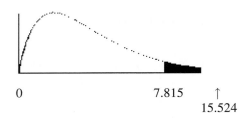

$$H = \frac{12}{26(26+1)}\left(\frac{21^2}{6} + \frac{96^2}{6} + \frac{138^2}{7} + \frac{96^2}{7}\right)$$

$$- 3(26+1)$$

H = 15.524

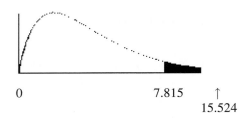

0                7.815        ↑
                            15.524

Reject the null hypothesis. There is enough evidence to conclude a difference in beach temperatures.

Without the Southern Pacific:

C. V. = 5.991 at $\alpha = 0.05$

$$H = \frac{12}{20(20+1)}\left(\frac{60^2}{6} + \frac{96^2}{7} + \frac{54^2}{7}\right)$$

$$- 3(20+1)$$

H = 3.661

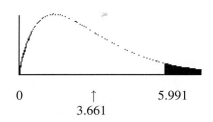

0            ↑            5.991
          3.661

Do not reject the null hypothesis. There is not enough evidence to conclude a difference in the temperatures. The conclusion is not the same without the Southern Pacific temperatures.

11.

| Last Year | $R_1$ | This Year | $R_2$ | $R_1$- $R_2$ | $d^2$ |
|---|---|---|---|---|---|
| 28.9 | 10 | 26.6 | 10 | 0 | 0 |
| 26.4 | 9 | 20.5 | 9 | 0 | 0 |
| 20.8 | 6 | 20.2 | 8 | − 2 | 4 |
| 25.0 | 8 | 19.1 | 7 | 1 | 1 |
| 21.0 | 7 | 18.9 | 6 | 1 | 1 |
| 19.2 | 5 | 17.8 | 5 | 0 | 0 |
| 13.7 | 1 | 16.8 | 4 | − 3 | 9 |
| 18.8 | 4 | 16.7 | 3 | 1 | 1 |
| 16.8 | 3 | 16.0 | 2 | 1 | 1 |
| 15.3 | 2 | 15.8 | 1 | 1 | 1 |
| | | | | $\sum d^2 =$ | 18 |

$$r_s = 1 - \frac{6\sum d^2}{n(n^2-1)} = 1 - \frac{6(18)}{10(99)} = 0.891$$

$H_0$: $\rho = 0$
$H_1$: $\rho \neq 0$

C. V. = $\pm 0.648$

Reject the null hypothesis. There is a significant relationship in the average number of people who are watching the television shows for both years.

13.
$H_0$: The grades of students who finish the exam occur at random.
$H_1$: The null hypothesis is not true.

The median grade is 73. Using A = above and B = below, the runs are:

AAAA B AAAA BBBB AAAAA BB A BBBBBBB

Since there are eight runs and this does not fall between 9 and 21, the null hypothesis is rejected. The grades do not occur at random.

CHAPTER 13 QUIZ

1. False
2. False, they are less sensitive.
3. True
4. True
5. a.
6. c.
7. d.
8. b.
9. non-parametric
10. nominal, ordinal
11. sign
12. sensitive

13. $H_0$: Median = 300 (claim)
$H_1$: Median $\neq$ 300
There are seven + signs. Do not reject since 7 is greater than the critical value 5. There is not enough evidence to reject the claim that the median is 300.

14. $H_0$: Median = 1200 (claim)
$H_1$: Median $\neq$ 1200
There are ten − signs. Do not reject since 10 is greater than the critical value 6. There is not enough evidence to reject the claim that the median is 1,200.

15. $H_0$: There will be no change in the weight of the turkeys after the special diet.
$H_1$: The turkeys will weigh more after the special diet. (claim)
There is one + sign. Reject the null hypothesis. There is enough evidence to support the claim that the turkeys gained weight on the special diet.

16. $H_0$: The distributions are the same.
$H_1$: The distributions are different. (claim)
C. V. = $\pm 1.96$
z = $-0.05$
Do not reject the null hypothesis. There is not enough evidence to reject the claim that the distributions are the same.

17. $H_0$: The distributions are the same.
$H_1$: The distributions are different. (claim)
C. V. = $\pm 1.65$
z = $-0.14$
Do not reject the null hypothesis. There is not enough evidence to support the claim that the distributions are different.

18. $H_0$: There is no difference in the GPAs before and after the workshop.
$H_1$: There is a difference in GPAs before and after the workshop. (claim)
C. .V. = 2    Test statistic = 0
Reject the null hypothesis. There is enough evidence to support the claim that there is a difference in the GPAs of the students.

19. $H_0$: There is no difference in the breaking strengths of the tapes.
$H_1$: There is a difference in breaking strengths. (claim)
H = 29.25
$\chi^2 = 5.991$
Reject the null hypothesis. There is enough evidence to support the claim that there is a difference in the breaking strengths of the tapes.

20. $H_0$: There is no difference in reaction times.
$H_1$: There is a difference in reaction times. (claim)
H = 6.9
0.025 < P-value < 0.05   (0.032)
Reject the null hypothesis. There is enough evidence to support the claim that there is a difference in the reaction times of the monkeys.

21. $H_0$: $\rho = 0$
$H_1$: $\rho \neq 0$
C. V. = $\pm 0.600$
$r_s = 0.683$
Reject the null hypothesis. There is a significant relationship between the drug prices.

22. $H_0$: $\rho = 0$
$H_1$: $\rho \neq 0$
$r_s = 0.943$
C. V. = $\pm 0.829$
Reject the null hypothesis. There is a significant relationship between the amount of money spent on Head Start and the number of students enrolled in the program.

23. $H_0$: The gender of babies occurs at random.
$H_1$: The null hypothesis is false.
$\alpha = 0.05$    C. V. = 8, 19
There are 10 runs, which is between 8 and 19. Do not reject the null hypothesis. There is not enough evidence to reject the null hypothesis that the gender occurs at random.

24. $H_0$: There is no difference in output ratings.
$H_1$: There is a difference in output ratings after reconditioning. (claim)
$\alpha = 0.05$    n = 9    C. V. = 6
test statistic = 0
Do not reject the null hypothesis. There is not enough evidence to support the claim that there is a difference in output ratings before and after reconditioning.

25. $H_0$: The numbers occur at random
$H_1$: The null hypothesis is false.
$\alpha = 0.05$    C. V. = 9, 21
The median number is 538.
There are 16 runs, reading from left to right, and since this is between 9 and 21, the null hypothesis is not rejected. There is not enough evidence to reject the null hypothesis that the numbers occur at random.

EXERCISE SET 14-1

1.
Random, systematic, stratified, cluster.

3.
A sample must be randomly selected.

5.
Talking to people on the street, calling people on the phone, and asking one's friends are three incorrect ways of obtaining a sample.

7.
Random sampling has the advantage that each unit of the population has an equal chance of being selected. One disadvantage is that the units of the population must be numbered, and if the population is large this could be somewhat time consuming.

9.
An advantage of stratified sampling is that it ensures representation for the groups used in stratification; however, it is virtually impossible to stratify the population so that all groups could be represented.

11 through 15.
Answers will vary.

EXERCISE SET 14-2

1.
This is a biased question. Change the question to read: "Will you vote for John Doe or Bill Jones for class president?"

3. This is a biased question. Change the question to read: "Should banks charge a fee to balance their customers' checkbooks?"

5.
This question has confusing wording. Change the question to read: "How many hours did you study for this exam?"

7.
This question has confusing wording. Change the question to read: "If a plane were to crash on the border of New York and New Jersey, where should the victims be buried?"

9.
Answers will vary.

EXERCISE SET 14-3

1.
Simulation involves setting up probability experiments that mimic the behavior of real life events.

3.
John Van Neumann and Stanislaw Ulam.

5.
The steps are:
1. List all possible outcomes.
2. Determine the probability of each outcome.
3. Set up a correspondence between the outcomes and the random numbers.
4. Conduct the experiment using random numbers.
5. Repeat the experiment and tally the outcomes.
6. Compute any statistics and state the conclusions.

7.
When the repetitions increase there is a higher probability that the simulation will yield more precise answers.

9.
Use two-digit random numbers 00 through 74 to represent the users; 75 through 99 to represent nonusers.

11.
Use one-digit random numbers 1 through 3 to represent a hit. Use three digits, 000 through 270 to represent a hit.

13.
Let an odd number represent heads and an even number represent tails. Then each person selects a digit at random.

15 through 23.
Answers will vary.

REVIEW EXERCISES - CHAPTER 15

1 - 3.
Answers will vary.

**5 - 7.**
Answers will vary. The range of the data is fairly narrow, and the data are not ranked in any way. Regardless of the method of sampling used, the sample means should be about the same.

**9.**
Use one-digit random numbers 1 through 4 to represent a strikeout and 5 through 9 and 0 to represent anything other than a strikeout.

**11.**
In this case, a one-digit random number is selected. Numbers 1 through 6 represent the numbers on the face. Ignore 7, 8, 9, and 0 and select another number.

**13.**
Let the digits 1 − 3 represent "rock"
Let the digits 4 − 6 represent "paper"
Let the digits 7 − 9 represent "scissors"
Omit 0.

**15 - 17.**
Answers will vary.

**19.**
This is a biased question. Change the question to read: "Have you ever driven through a red light?"

**21.**
This is a double-barreled question. Change the question to read: "Do you think all automobiles should have heavy-duty bumpers?"

**CHAPTER 14 QUIZ**

1. True
2. True
3. False, only random numbers generated by a random number table are random.
4. True
5. a.
6. c.
7. c.
8. larger
9. biased
10. cluster
11 through 14. Answers will vary.

**15.** Use two-digit random numbers: 01 through 45 constitute a win. Any other two-digit number means the player loses.

**16.** Use two-digit random numbers: 01 - 05 constitute a cancellation. Any other two-digit random number means the person shows up.

**17.** Use two-digit random numbers 01 through 13 to represent the 13 cards in hearts. The random numbers 14 through 26 represent the 13 cards in diamonds. The random numbers 27 through 39 represent the 13 spades, and 40 through 52 represent the 13 clubs. Any number over 52 is ignored.

**18.** Use two-digit random numbers to represent the spots on the face of the dice. Ignore any two-digit random numbers with 7, 8, 9, or 0. For cards, use two-digit random numbers between 01 and 13.

**19.** Use two-digit random numbers. The first digit represents the first player, and the second digit represents the second player. If both numbers are odd or even, player 1 wins. If a digit is odd and the other digit is even, player 2 wins.

20 through 24. Answers will vary.

A-1

A-1. $9! = 9 \cdot 8 \cdot 7 \cdot 6 \cdot 5 \cdot 4 \cdot 3 \cdot 2 \cdot 1 = 362,880$

A-2. $7! = 7 \cdot 6 \cdot 5 \cdot 4 \cdot 3 \cdot 2 \cdot 1 = 5040$

A-3. $5! = 5 \cdot 4 \cdot 3 \cdot 2 \cdot 1 = 120$

A-4. $0! = 1$

A-5. $1! = 1$

A-6. $3! = 3 \cdot 2 \cdot 1 = 6$

A-7. $\frac{12!}{9!} = \frac{12 \cdot 11 \cdot 10 \cdot 9!}{9!} = 1320$

A-8. $\frac{10!}{2!} = \frac{10 \cdot 9 \cdot 8 \cdot 7 \cdot 6 \cdot 5 \cdot 4 \cdot 3 \cdot 2!}{2!}$

$= 1,814,400$

A-9. $\frac{5!}{3!} = \frac{5 \cdot 4 \cdot 3!}{3!} = 20$

A-10. $\frac{11!}{7!} = \frac{11 \cdot 10 \cdot 9 \cdot 8 \cdot 7!}{7!} = 7920$

A-11. $\frac{9!}{(4!)(5!)} = \frac{9 \cdot 8 \cdot 7 \cdot 6 \cdot 5!}{4 \cdot 3 \cdot 2 \cdot 1 \cdot 5!} = 126$

A-12. $\frac{10!}{(7!)(3!)} = \frac{10 \cdot 9 \cdot 8 \cdot 7!}{3 \cdot 2 \cdot 1 \cdot 7!} = 120$

A-13. $\frac{8!}{4!4!} = \frac{8 \cdot 7 \cdot 6 \cdot 5 \cdot 4!}{4 \cdot 3 \cdot 2 \cdot 1 \cdot 4!} = 70$

A-14. $\frac{15!}{12!3!} = \frac{15 \cdot 14 \cdot 13 \cdot 12!}{3 \cdot 2 \cdot 1 \cdot 12!} = 455$

A-15. $\frac{10!}{(10!)(0!)} = \frac{10!}{10! \cdot 1} = 1$

A-16. $\frac{5!}{3!2!1!} = \frac{5 \cdot 4 \cdot 3!}{3! \cdot 2 \cdot 1 \cdot 1} = 10$

A-17. $\frac{8!}{3!3!2!} = \frac{8 \cdot 7 \cdot 6 \cdot 5 \cdot 4 \cdot 3!}{3! \cdot 3 \cdot 2 \cdot 1 \cdot 2 \cdot 1} = 560$

A-18. $\frac{11!}{7!2!2!} = \frac{11 \cdot 10 \cdot 9 \cdot 8 \cdot 7!}{7! \cdot 2 \cdot 1 \cdot 2 \cdot 1} = 1980$

A-19. $\frac{10!}{3!2!5!} = \frac{10 \cdot 9 \cdot 8 \cdot 7 \cdot 6 \cdot 5!}{3 \cdot 2 \cdot 1 \cdot 2 \cdot 1 \cdot 5!} = 2520$

A-20. $\frac{6!}{2!2!2!} = \frac{6 \cdot 5 \cdot 4 \cdot 3 \cdot 2!}{2 \cdot 1 \cdot 2 \cdot 1 \cdot 2!} = 90$

A-2

A-21.

| X | $X^2$ | $X - \overline{X}$ | $(X - \overline{X})^2$ |
|---|---|---|---|
| 9 | 81 | $-3.1$ | 9.61 |
| 17 | 289 | 4.9 | 24.01 |
| 32 | 1024 | 19.9 | 396.01 |
| 16 | 256 | 3.9 | 15.21 |
| 8 | 64 | $-4.1$ | 16.81 |
| 2 | 4 | $-10.1$ | 102.01 |
| 9 | 81 | $-3.1$ | 9.61 |
| 7 | 49 | $-5.1$ | 26.01 |
| 3 | 9 | $-9.1$ | 82.81 |
| <u>18</u> | <u>324</u> | <u>5.9</u> | <u>34.81</u> |
| 121 | 2181 | | 716.9 |

$\sum X = 121$  $\overline{X} = \frac{121}{10} = 12.1$  $\sum X^2 = 2181$

$(\sum X)^2 = 121^2 = 14641$  $\sum(X - \overline{X})^2 = 716.9$

A-22.

| X | $X^2$ | $X - \overline{X}$ | $(X - \overline{X})^2$ |
|---|---|---|---|
| 4 | 16 | $-3$ | 9 |
| 12 | 144 | 5 | 25 |
| 9 | 81 | 2 | 4 |
| 13 | 169 | 6 | 36 |
| 0 | 0 | $-7$ | 49 |
| 6 | 36 | $-1$ | 1 |
| 2 | 4 | $-5$ | 25 |
| <u>10</u> | <u>100</u> | <u>3</u> | <u>9</u> |
| 56 | 550 | | 158 |

$\sum X = 56$  $\overline{X} = \frac{56}{8} = 7$  $\sum X^2 = 550$

$(\sum X)^2 = 56^2 = 3136$  $\sum(X - \overline{X})^2 = 158$

A-23.

| X | $X^2$ | $X - \overline{X}$ | $(X - \overline{X})^2$ |
|---|---|---|---|
| 5 | 25 | $-1.4$ | 1.96 |
| 12 | 144 | 5.6 | 31.36 |
| 8 | 64 | 1.6 | 2.56 |
| 3 | 9 | $-3.4$ | 11.56 |
| <u>4</u> | <u>16</u> | <u>$-2.4$</u> | <u>5.76</u> |
| 32 | 258 | | 53.20 |

$\sum X = 32$  $\overline{X} = \frac{32}{5} = 6.4$  $\sum X^2 = 258$

$(\sum X)^2 = 32^2 = 1024$  $\sum(X - \overline{X})^2 = 53.2$

A-24.

| X | $X^2$ | $X - \overline{X}$ | $(X - \overline{X})^2$ |
|---|---|---|---|
| 6 | 36 | $- 12.75$ | 163.5625 |
| 2 | 4 | $- 16.75$ | 280.5625 |
| 18 | 324 | $- 0.75$ | 0.5625 |
| 30 | 900 | 11.25 | 126.5625 |
| 31 | 961 | 12.25 | 150.0625 |
| 42 | 1764 | 23.25 | 540.5625 |
| 16 | 256 | $- 2.75$ | 7.5625 |
| 5 | 25 | $- 13.75$ | 189.0625 |
| 150 | 4270 | | 1457.5000 |

$\sum X = 150 \quad \overline{X} = \frac{150}{8} = 18.75 \quad \sum X^2 = 4270$

$(\sum X)^2 = 150^2 = 22500 \quad \sum(X-\overline{X})^2 = 1457.5$

A-25.

| X | $X^2$ | $X - \overline{X}$ | $(X - \overline{X})^2$ |
|---|---|---|---|
| 80 | 6400 | 14.4 | 207.36 |
| 76 | 5776 | 10.4 | 108.16 |
| 42 | 1764 | $- 23.6$ | 556.96 |
| 53 | 2809 | $- 12.6$ | 158.76 |
| 77 | 5929 | 11.4 | 129.96 |
| 328 | 22678 | | 1161.20 |

$\sum X = 328 \quad \overline{X} = \frac{328}{5} = 65.6 \quad \sum X^2 = 22678$

$(\sum X)^2 = 328^2 = 107584 \quad \sum(X-\overline{X})^2 = 1161.2$

A-26.

| X | $X^2$ | $X - \overline{X}$ | $(X - \overline{X})^2$ |
|---|---|---|---|
| 123 | 15129 | $- 15.17$ | 230.1289 |
| 132 | 17424 | $- 6.17$ | 38.0689 |
| 216 | 46656 | 77.83 | 6057.5089 |
| 98 | 9604 | $- 40.17$ | 1613.6289 |
| 146 | 21316 | 7.83 | 61.3089 |
| 114 | 12996 | $- 24.17$ | 584.1889 |
| 829 | 123125 | | 8584.8334 |

$\sum X = 829 \quad \overline{X} = \frac{829}{6} = 138.17$

$\sum X^2 = 123125 \quad (\sum X)^2 = 829^2 = 687241$

$\sum(X-\overline{X})^2 = 8584.8334$

A-27.

| X | $X^2$ | $X - \overline{X}$ | $(X - \overline{X})^2$ |
|---|---|---|---|
| 53 | 2809 | $- 16.3$ | 265.69 |
| 72 | 5184 | 2.7 | 7.29 |
| 81 | 6561 | 11.7 | 136.89 |
| 42 | 1764 | $- 27.3$ | 745.29 |
| 63 | 3969 | $- 6.3$ | 39.69 |
| 71 | 5041 | 1.7 | 2.89 |
| 73 | 5329 | 3.7 | 13.69 |
| 85 | 7225 | 15.7 | 246.49 |
| 98 | 9604 | 28.7 | 823.69 |
| 55 | 3025 | $- 14.3$ | 204.49 |
| 693 | 50511 | | 2486.10 |

$\sum X = 693 \quad \overline{X} = \frac{693}{10} = 69.3 \quad \sum X^2 = 50511$

$(\sum X)^2 = 693^2 = 480249 \quad \sum(X-\overline{X})^2 = 2486.1$

A-28.

| X | $X^2$ | $X - \overline{X}$ | $(X - \overline{X})^2$ |
|---|---|---|---|
| 43 | 1849 | $- 38.8$ | 1505.44 |
| 32 | 1024 | $- 49.8$ | 2480.04 |
| 116 | 13456 | 34.2 | 1169.64 |
| 98 | 9604 | 16.2 | 262.44 |
| 120 | 14400 | 38.2 | 1459.24 |
| 409 | 40333 | | 6876.80 |

$\sum X = 409 \quad \overline{X} = \frac{409}{5} = 81.8 \quad \sum X^2 = 40333$

$(\sum X)^2 = 409^2 = 167281 \quad \sum(X-\overline{X})^2 = 6876.8$

A-29.

| X | $X^2$ | $X - \overline{X}$ | $(X - \overline{X})^2$ |
|---|---|---|---|
| 12 | 144 | $- 41$ | 1681 |
| 52 | 2704 | $- 1$ | 1 |
| 36 | 1296 | $- 17$ | 289 |
| 81 | 6561 | 28 | 784 |
| 63 | 3969 | 10 | 100 |
| 74 | 5476 | 21 | 441 |
| 318 | 20150 | | 3296 |

$\sum X = 318 \quad \overline{X} = \frac{318}{6} = 53 \quad \sum X^2 = 20150$

$(\sum X)^2 = 318^2 = 101124 \quad \sum(X-\overline{X})^2 = 3296$

A-30.

| X | $X^2$ | $X - \overline{X}$ | $(X - \overline{X})^2$ |
|---|---|---|---|
| − 9 | 81 | − 5.67 | 32.1489 |
| − 12 | 144 | − 8.67 | 75.1689 |
| 18 | 324 | 21.33 | 454.9689 |
| 0 | 0 | 3.33 | 11.0889 |
| − 2 | 4 | 1.33 | 1.7689 |
| − 15 | 225 | − 11.67 | 136.1889 |
| − 20 | 778 | | 711.3334 |

$\overline{X} = \frac{-20}{6} = -3.33 \qquad (\sum X)^2 = -20^2 = 400$

A-33.

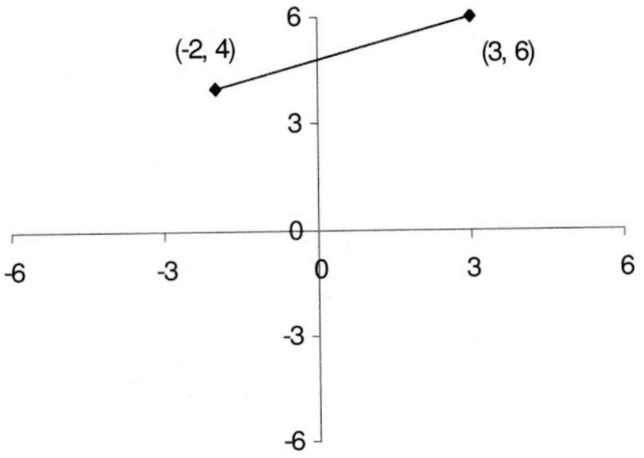


A-3

A-31.

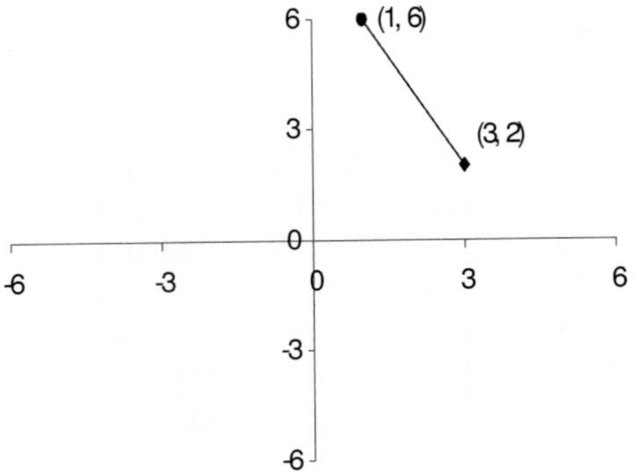


A-34.

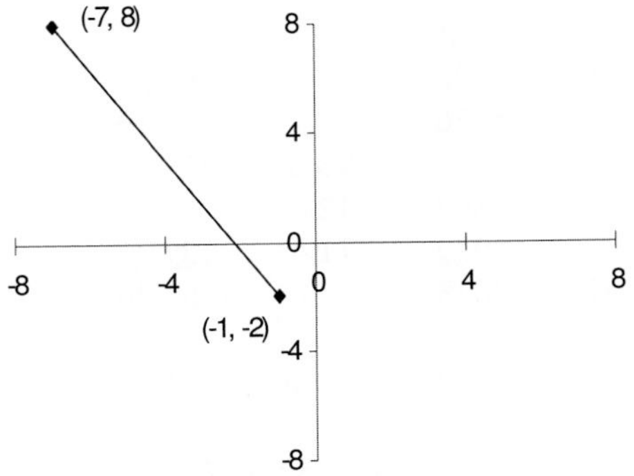


A-32.

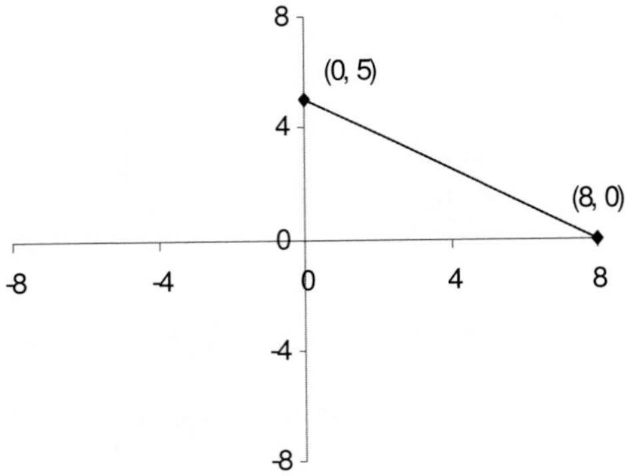


A-35.

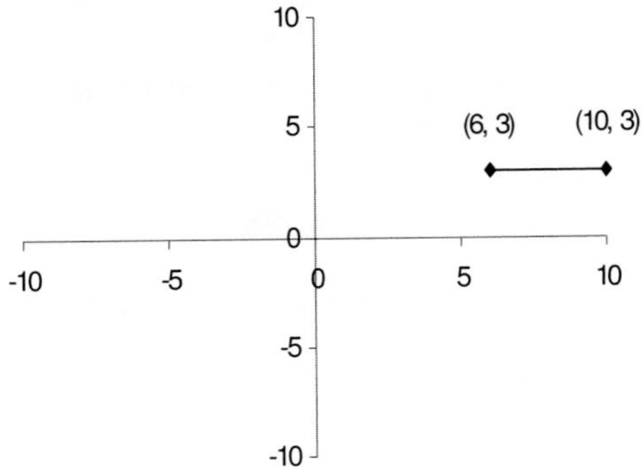

A-36.

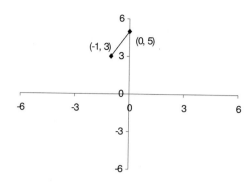

Two points are: (0, 5) and (-1, 3).

A-37.

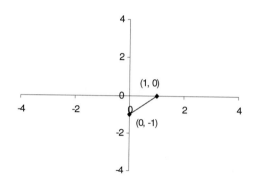

Two points are: (1, 0) and (0, -1).

A-38.

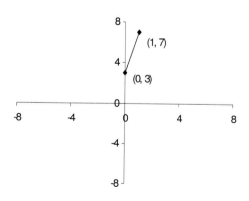

Two points are: (1, 7) and (0, 3).

A-39

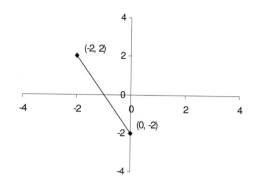

Two points are: (-2, 2) and (0, -2).

A-40.

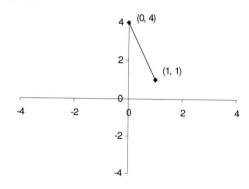

Two points are: (1, 1) and (0, 4)

# Notes

# Notes

# Notes